A MATHEMATICS
UNIFIES THE

UNIVERSE

Bardolf & Company

A MATHEMATICS UNIFIES THE UNIVERSE

ISBN 978-1-938842-47-4
Copyright © 2020 by Hector Estepan

Published by Bardolf & Company
www.bardolfandcompany.com

Cover design by Shaw Creative
www.shawcreativegroup.com

I dedicate this book to

Josefa Estepan

my mother who always challenged me with puzzles,
encouraged me to always try to understand
whatever was around me,
to get up if I failed and try again,
and to never give up in things I believe in,

Carol Magett

who always believed my unification quest and spent
many hours discussing unification with me,

Rosina Ramos

my aunt who brought my family
to this country she loved so much
and gave me the opportunity and exposure to
its outstanding education,

and to

Lowell Hawthorne

my dear friend who was extremely encouraging
and inspirational in my quest.

Give me a lever long enough and a fulcrum on which to place it, and I shall move the world.

—Archimedes

A MATHEMATICS
UNIFIES THE
UNIVERSE

HECTOR ESTEPAN

Bardolf & Company
Sarasota, Florida

INTRODUCTION

In 1992, I tried to figure out how to best convey Einstein' s General Relativity theory to my 16-year-old son. It occurred to me that there should be an aether between celestial bodies, a medium that carried light waves. But the 1887 Michelson-Morley experiment failed to detect any such medium, leading to the conclusion that there was no aether. Subsequent scientific discoveries, including Einstein's special relativity, were all based on the non-aether notion.

Although the original experiment has been confirmed numerous times, I decided to perform it again to verify the results. On a hunch, I also looked at the effect gravity had on the measured light and found that the ratio of the Sun's gravitational force on the Moon was 2.19 times greater than the Earth's! I was amazed. How could that be? Just to be sure, I repeated the calculation several times, using different sources for the values in the ratio and got the same result. Still incredulous, I asked a medical resident with a background in physics to calculate the ratio, and he came up with 2.19, too. Accepting that the Sun attracts the Moon with a force 2.19 times stronger than the Earth leads to the question: If gravity is the cause of the orbit, why does the Moon orbit the Earth?

My quest for an answer culminated in a book I wrote in 2012, *A Unification Theory of the Universe.* I used conventional mathematics which has, as the foundation of all mathematical functions, Euclid's point "that which has no part," meaning it is an imaginary construct with no dimensions. But that yields to abstract equations divorced from reality. For example, it is possible to have an infinite number of points in the same location an inconceivable notion in our material universe.

By 2015, I was having difficulties explaining reality using Euclid's point and asked myself: What if the mathematical point were real, such as a sphere having a non-zero radius? A point with three dimensions, i.e., volume, requires a different approach. For example, it requires having only one point in any particular location.

What I found was that using a three-dimensional mathematical point opened magnificent and splendorous discoveries in the process of creating a beautiful, logical universe that conforms more closely to reality than the Euclidean model. Any real moving object has a vector derivative (possessing both magnitude and direction; unlike a scalar, which has only magnitude), and provides an explanation for my original question regarding the Sun's attractive force. In space, an object can take an infinite number of orbits, so it settles on the smallest possible— the shortest orbital distance and minimal orbital time—in the case of the Moon, around the Earth.

This law was discovered by Johann Kepler in 1618, but no one paid attention to its implications. Newton said that by using his own equations, he could generate Kepler's third law, but that didn't explain anything. Kepler argued that what determines how fast an object goes around another object is a matter of spin. He found that the orbital time to the second power is proportional to the radial distance to the third power. When Kepler looked at the moons orbiting Jupiter, he posited that although he could not see the spin of the planet—telescopes at the time were not strong enough—it must have spin. And he was right.

There are other implications of using a three-dimensional mathematical point to create a realistic mathematical universe.

One of them has a profound affect on our notion of classical physics, as follows: Time is dependent on the dimensions of the moving object. All moving objects have to spin about the axis of motion. In turn, motion creates mass. Motion creates charge. Motion creates Maxwell's electromagnetic laws. Motion creates heat. Finally, motion requires an aether.

Another consequence relates to atomic physics: Two colliding mathematical points lead to the formation of different electrons, one type of neutron, and large number of different protons, which create black holes, stars, planets, and moons. The electrons and protons, without any neutrons, make the elements in the periodic table. Finally, in an aether, spinning objects engender gravity, which, according to Einstein, is equivalent to an acceleration.

Still another is that the mathematical point is the photon in the cosmos and generates cosmic microwave background radiation (CMB). A photon also has a charge! And a mass!

There are other general and detailed implications, as the reader will discover.

Because the math can get complicated, I have included explanations for general (lay) readers along the way so they can follow the arguments.

This book, then, is the road to the final frontier in this incredible, deterministic universe. The concepts articulated here have profound implications for the Big Bang theory, space travel, and our understanding of the future of our planet.

Enjoy the book and above all else, be critical. Although I am convinced that the real, three-dimensional mathematical point is the necessary concept to decipher the way the universe works, in the spirit of logic and scientific discovery, I welcome any comments, arguments, and refinement that can lead to further discoveries.

Hector Estepan
December 2019

AXIOMS and THEOREMS

Let me begin with two foundational axioms:

AXIOM 1. *The infinite volume of the mathematical universe V_U is the sum of two mutually exclusive volumes— the space volume V_S and the object volume V:*

$$V_U = V_S + V.$$

There is either space volume or object volume. The intersection of the space volume and the object volume is null.

AXIOM 2. *The vector $x\mathbf{i}_x + y\mathbf{i}_y + z\mathbf{i}_z$ determines every location in the universe.*

The object has location vectors, and space has location vectors.

TIME

Existing in the mathematical universe, time has no beginning and no end. It is a forward counter measuring equal distances from prior, during, and after the existence of object volume, and creates volume motion through the mathematical equation

$$V/f(t),$$

where $f(t) \neq 0$.

AXIOM 3. *Two distinct identical object volumes cannot occupy the same space and time.*

This axiom marks the defining difference between this mathematical universe and the Euclidean world.

Axiom 1 and Axiom 2 yield:

THEOREM 1. *Every object volume V has the vector*

$$Xi_x + Yi_y + Zi_z$$

where $X = f_1(x) \neq 0, Y = f_2(y) \neq 0,$ and $Z = f_3(z) \neq 0$.

If any X, Y, or Z is zero, then the object is not a volume, and the object is not real.

THEOREM 2. *A volume in motion must have the moving vector*

$$(Xi_x + Yi_y + Zi_z)/f(t)$$

where t is time, and $f(t) \neq 0$.

Proof: A volume has the vector $Xi_x + Yi_y + Zi_z$. A moving volume is $V/f(t)$. Therefore, everything in, and on the volume is divisible by $f(t)$.

THEOREM 3. *In the universe, any object volume in motion causes the space volume to be in motion.*

Proof: V_S and V are mutually exclusive. An object in motion is V/t. The object's volume V displaces V_S, and the moving object volume, V/t, imparts the same speed to an equal volume of space. If the volume and speed of the moving space is different from the object's volume and speed, then the space volume and the object's volume are not mutually exclusive. As the object volume moves, its coordinates are moving, causing movement of an equal space volume, which must have moving space coordinates. Therefore, all space coordinates change with the same rate as the object's coordinate rate of change.

Theorem 3 states that any movement in the universe transmits this movement to every location in the universe and causes every location in the universe to change its coordinates. This is action at a distance. The creation of a point with volume and motion explains action at a distance with no need to invoke any force or energy field as a cause. Theorem 3 is the action-at-a-distance theorem. Object movement transmits its movement to the entire universe.

Corollary 1. *An object's volume immersed in fluid displaces a fluid volume equal to the object's immersed volume.*

In space, a three-dimensional object displaces a space volume equal to the object's volume, and an object moving at constant speed v_o causes the equal space volume to move with speed v_o. An object in motion has a changing coordinate rate and transmits this rate to all objects in the universe, but since the space must move to make room for the moving volume, the space coordinate's rate of change must be identical to the object's coordinate's rate of change. Therefore, the message of an object's volume motion cannot be any faster or slower than the object's motion. In this mathematical universe, there is no question that action at a distance exist.

The lay reader may be overwhelmed with the mathematical symbols, which convey familiar concepts such as volume, location and movement. The latter is distance covered over time. Movement needs both distance and time. It is impossible without them.

DISPLACEMENT

How does volume interact in various media?

Tossing a stone into a pond of infinite radius generates an infinite number of waves. With a constant volume V, the stone generates discrete circles with discrete increasing radius r, and discrete decreasing height h. The radius r, and the height h can be easily seen on the pond's surface. The pond's surface wave only appears when the stone displaces a volume of water equal to the stone's volume, but a floating object displaces its immersed volume, which causes the circular wave.

A stone displaces its total volume in the water. A floating object displaces part of its volume and not its entire volume. No matter the object shape, if it sinks, then its volume is the displaced water volume. On the other hand, if the object floats, the weight of the displaced water is the object's weight.

The displaced volume causes the waves. No fraction of the displaced volume causes waves. The shape of the displaced volume is spherical. Mathematically, the sphere has the smallest surface area enclosing a given volume, and surrounds the largest volume among all closed surfaces. In the mathematical universe, the sphere has the minimal surface enclosing the greatest volume. The sphere is unique in the mathematical and real universes. No other three-dimensional shape satisfies the constraint of enclosing a given volume with the minimal surface area. This is how nature transmits information to the universe—its process utilizes the most efficient and economical way: By converting the volume to a spherical volume, moving the distance of the spherical radius just created, and displaces the volume so that at each discrete r, the maximal crest height is at the discrete r distance. How this process comes about is a mystery. One visible demonstration occurs when introducing motion to a certain amount of soap water volume, which creates a sphere as the soap bubble displaces the atmospheric medium. Another is a raindrop falling on the pond, which results in a process that converts the raindrop volume into a sphere that transmits the raindrop volume to the entire pond.

The wave creating process converts any volume, and in this case, the stone's volume, into the spherical volume $V = (4/3)\pi r^3$. The spherical volume moves the discrete distance r to inform the water molecules in the spherical volume between r and $2r$ that the constant volume, $(4/3)\pi r^3$, is distributing itself to the $2r$ sphere such that, at location $2r$, h_2 is maximal. The mathematical function for h as a function of r is complex, but h_2 is maximal at r_2. Traveling the specific distance r, the volume $(4/3)\pi r^3$ informs the water molecules in the spherical volume between $2r$ and $3r$ that the constant volume, $(4/3)\pi r^3$, is distributing itself so that at $3r$, h_3 reaches its maximum. In the infinite pond, the process goes on in perpetuity, and takes place irrespective of the medium, whether it is water, oil, molasses, atmosphere, or space.

The constant speed of the moving volume $(4/3)\pi r^3$ eventually reaches the pond's edge, with the constant volume speed creating continuity, and the volume distribution creating the quantum. The continuity function is the constant speed of the volume informing all the water molecules in the pond that the discrete volume is in the pond, and it informs the pond's water molecules in packets of the displaced volume $(4/3)\pi r^3$ at specific locations $r, 2r, 3r,..., nr$.

For the lay reader, the idea is that any volume immersed in water informs the rest of the water molecules as to how and when the water molecules distribute themselves. This can be seen when a visible wave travels on the surface of a pond. The transmission of the new volume is not an instantaneous process, however, as Isaac Newton believed it to be. It takes time for the wave to travel to the edge of the pond, as Einstein has shown.

This raises the question: How fast does volume travel to inform the pond?

The volume equation is

$$V = \pi r^2 h,$$

or

$$h = V/\pi r^2.$$

The derivative of h with respect to time is

$$dh/dt = -(2V/\pi r^3)dr/dt$$

with a radial expansion rate

$$dr/dt = -(\pi r^3/2V)dh/dt.$$

As the distance from the radius origin increases, the radial expansive rate, dr/dt, increases. Equating the pond to the mathematical universe, shows that the ripples expand at a faster rate the farther the observation is from the initial disturbance. Because three-dimensional objects have a volume, their volume movement creates a seemingly expanding universe, but the moving immersed object does not expand the volume of the universe. Rather the object's motion rearranges the universal volume, giving the impression of an expanding universe. The prevalent belief that the universe is expanding is not correct. We are merely watching waves in motion.

Object movement is the quantum transmitting method to the entire mathematical universe. Through the moving volume, it eventually informs every single location in the mathematical universe. The transmission distance is immaterial, with no difference between the infinitesimal distance in atoms to the parsec distance in space. The process is not continuous but more discrete, but dealing with transmission rate close to the speed of light the process is almost continuous and the above differentiation may be close to the answer.

THE INVERSE SQUARE LAW

The observed wave $h = V/\pi r^2$ is inversely related to the radius squared. Thus, in the mathematical universe, anything moving creates a disturbance, h, which is inversely related to the distance squared from the initial disturbance site. Moving masses and moving charges are moving volumes, which create disturbance h, giving rise to the inverse square law. Through displacement, the moving volume holds the key to Feynman's perplexing question, which asks how the gravitational equation $F = GMm/r^2$ is related to the electromagnetic equation $F = q_1q_2/4\pi\varepsilon_o r^2$.

Although, the two equations describe different phenomena, they have almost the same form. Is it possible that they have the same origin? I believe so. The common determinant is volume displacement. That is the unifying link between the two equations.

Because the mathematical point is three-dimensional, i.e. has volume, the mathematical universe is realistic. Axioms 1 and

3 offer an alternative view from the Euclidean universe. Axiom 3 describes nothing more complicated than an object entering a bathtub partially filled with water, which creates an increase water height. With two-dimensional mathematical points, which have no volume, such displacement is impossible—in any medium.

For the lay reader, the universe is a volume composed of two mutually exclusive volumes: the volumes of real things, and everything that does not have a boundary, i.e., space. Objects like planets, stars, comets are real. The space between is boundless.

For most people, the vector concept may be too abstract, but vectors are all around us. Just look at the corner of any room and you will see length, width, and height. Everything that is real can be expressed in length, width, and height, and there are no exceptions. The vector equation posits that there is a length, $X\mathbf{i}_x$, a width $Y\mathbf{i}_y$, and a height $Z\mathbf{i}_z$. Do not let the mathematical expressions for these concepts of length, width, and height scare you. Look at them as part of a language: When one says length, one says $X\mathbf{i}_x$. When one looks at the corner of a room, the corner of a box, the corner of a building, or the corner of a home, that corner's length, width and height can be expressed by the vector $X\mathbf{i}_x + Y\mathbf{i}_y + Z\mathbf{i}_z$.

Another concept we have to address is time. The word has use in a variety of disciplines, including philosophy and psychology. For our purposes, time is merely an interval between two locations. For instance, if the distance between two locations is 100 miles, and if an object is moving at 20 miles an hour, then the time distance between the two locations is 5 hours.

Since all real objects have a volume, then real objects must have length, width, and height; and an object in motion must be in motion along a direction composed of these three dimensions. While these directions can vary, along the direction of motion all of them must have the same speed; a dimension not having this speed does not belong to the object. From this notion, we may conclude that variable lengths along the direction of motion must have variable times. No experiment needs to be done to prove this because it logically follows. As simple as this concept is, it is not widely known, but it has breathtaking consequences.

One example is the twin paradox, a thought experiment based on Albert Einstein's discovery of special relativity. One twin remains on Earth and the other travels at a speed greater than the Earth's movement. Upon returning to Earth, the traveling twin is younger than the twin who stayed behind. Einstein's special relativity shows how motion causes the variance in space-time.

Movement of an object requires the concept of time. Without time there is no movement. An object in motion is a volume in motion, therefore everything in, and on the object must be in motion. The space outside a body in motion has coordinates; if a body moves, then its space coordinates must also move. It logically follows that if an object moves, the coordinates external to the moving object must move, therefore its motion becomes known to the entire universe. How fast the universe obtains the news depends on how fast that news travels throughout the universe. The moving concept, or action at a distance concept, is only determined by a volume moving. It has nothing to do with gravity, electromagnetism, sound, or light.

The last concept we need to discuss here is the displacement principle, which follows from the understanding that space and object volume are mutually exclusive. Consider that an object immersed in water displaces a volume of water equal to that of the object's volume. If we accept that $V = h\pi r^2$, expresses all volumes, then it follows that $h = V/\pi r^2$. This is, once again, the inverse square law discussed earlier, but it obtains not because of mass, charge or light intensity, but simply because of a volume displacement. That is to say, if mass has a volume, if charge has a volume, if light has a volume, and they are in motion, then each one of these three moving objects informs the universe of its movement according the inverse square law. It is volume movement that leads to the inverse square law, not the other way around.

THE MATHEMATICAL POINT

According to the inverse square law, an object in motion has line lengths in the direction of motion, with a moving area orthogonal (at a right angle) to the lines in the direction of motion. Based on the 3 Axioms and 3 Theorems, a real mathematical universe exists.

The mathematical point is a sphere having a radius

$$r_o \neq 0,$$

having a constant speed

$$v_o \neq 0,$$

having a volume

$$(4/3)\pi r o^3,$$

having a mass

$$m = 0$$

and having a charge

$$q = 0$$

These equations only define the mathematical point. What matters here is that the mathematical point is real and has a volume. Anything with a moving volume outlines a continuous path, which can be thought of as a mathematical function. Taking a pencil and drawing a line causes that pencil point to move and create that linear mathematical function. Thus, writing, painting, any motion outlines a mathematical function.

To create a real mathematical universe there must be motion—no motion, no universe.

CONTINUOUS MATHEMATICAL FUNCTION

The three-dimensional mathematical point, having the vector

$$(Xi_x + Y i_y + Zi_z)/t,$$

outlines a continuous mathematical function in the interval minus to positive infinity. Because the sphere is real and exists in the infinite interval, the sphere can neither be created or destroyed. This is a rather profound statement. If the moving sphere is to outline the mathematical function, the sphere is eternal.

This leads to:

THEOREM 4. *The mathematical point cannot be created or destroyed.*

Destroying the mathematical point destroys the mathematical function.

UNIFORM MATHEMATICAL FUNCTION

The function's uniformity depends on the constant speed v_o. In the mathematical universe, a varying v_o destroys uniformity.

The average speed in any given interval must be v_o, otherwise the uniform mathematical universe ceases to exist. This means that v_o is the constant speed that permeates the mathematical universe. Compare with the speed Einstein gives to the speed of light, then look at the discovery of the cosmic microwave background (CMB), composed of photons permeating the universe. Each one of these photons is moving at the constant speed of light. The similarity of the mathematical point in motion and photons is quite striking, but the current dogma is that the mathematical point is eternal, while a photon is thought to be ephemeral.

However, this disparity may not be true. If the photon is equivalent to the mathematical point, it cannot be created or destroyed either, in which case there needs to be a radical, conceptual change in the our understanding of the photon.

THE MATHEMATICAL FUNCTION

The mathematical function describes the path of the real spherical volume—a parabola, or any other mathematical function, but the space volume is not linear, parabolic, or based on a mathematical function. The path of the mathematical point dictates the space; the space des not dictate the path of the mathematical point.

THE MATHEMATICAL UNIVERSE

In the mathematical universe, the motion of mathematical points outlines the mathematical function and the space between the mathematical points. Every point in the mathematical universe is in motion, hence V/f(t) causes $V_S/f(t)$, but as $V_S/f(t)$ exists, other V_S moves; eventually all V_S moves. The entire mathematical universe moves. In the universe, object movement causes action at a distance. This phenomenon results in:

THEOREM 5. *All volumes are moving in the mathematical universe, and movement causes action at a distance.*

There is an infinite number of mathematical functions. For example, a set of parallel lines has an infinite number of parallel lines. Therefore, an infinite number of mathematical points exists. Space volume, V_S, and mathematical point volume, V, are mutually exclusive, therefore both cannot exist in the same time and place. In any given time and place, there is either space volume, V_S, or the mathematical point volume, V, but not both. Because the mathematical point is in a constant motion, $V/f(t)$, space volume, $V_S \neq 0$, allows the mathematical point volume to move. With no mathematical point movement, there is no mathematical function. From this notion, it follows that the space volume must be greater than object volume, otherwise there is no movement.

For the lay reader, it is very important to understand that the space volume must be greater than the object volume. That allows the sphere motion. In an atom, for the electron and the proton to be able to move, they must have more space than their relative size. The only way for that to happen is that the proton decreases in size. If the nuclear radius remains the same size, as the number of protons increases, each individual proton must decrease in size. As a result, the atom then is mostly empty space, just like the universe.

THE MOVING POINT

In a vacuum, a sphere of radius r_o moving with speed c is a moving volume $(4/3)\pi r_o^2 c$. Because there are a number of mathematical points in a given volume, or a density p, the ratio of the number of such volumes $n(4/3)\pi r_o^3$ to the given volume, a cubic centimeter, results in a dimensionless number. The moving point $(4/3)\pi r_o^2 c$ has a resistance, the product $p\pi r^2 C_d$, where C_d is the dimensionless drag coefficient, which depends on object shape and speed. For a high velocity sphere, C_d asymptotically approaches 0.47 as the sphere's velocity approaches infinity[1]. The resultant object speed is

$$v_o = (4/3)\pi r_o^2 c/(p\pi r_o^2 C_d).$$

Let $p = 389$, which leads to

$$v_o = (4/3)\pi r_o^2 c/(389\pi r_o^2 (0.47)) = c/137.1225,$$

where c is the speed of the sphere when there is no resistance. Letting C_d equal 0.46958012, which depends on the value of the finite speed c, results in

$$v_o = c/137.$$

The inverse of the prime number 137 is the number usually associated with the fine structure constant α. The questionable arbitrary prime number 389 is close to the calculated 400 value of cosmic microwave background radiation particles in a cubic centimeter. According to scientists, these photons are traveling at the speed of light, but the result is that the mathematical point is traveling at the speed of 1/137 of the speed of a mathematical point in a vacuum.

---

1   https://www.grc.nasa.gov/www/k-

For the lay reader, the number 137 appears in the atomic universe quite frequently. Arnold Sommerfeld, a German theoretical physicist, discovered it in 1916 when he measured the spectral lines in the structure of atoms and found that the variations are always a multiple of that number. But neither he, nor any other scientists have been able to explain why or discover a reason for its existence. The idea behind the mathematical manipulations is that it shows that given a number of mathematical points per unit volume, and the sphere with its moving shape yields a dimensionless number very close to 137 demonstrating that the described real mathematical point does have a place in reality.

CMB AND MATHEMATICAL POINT DENSITIES

It is generally agreed among scientists that the number of cosmic microwave background radiation, CMB, in a cubic centimeter is about 400. These particles (photons) are homogeneously spread throughout the cubic centimeter. That is what we observe, although we don't know how this point distribution happens.

With 389 homogeneously spread mathematical points in one cubic centimeter, then a one-centimeter line has the cube root of 389, or 7.2999893662 mathematical points. The distance between adjacent mathematical points is 0.1369882969 centimeter. Letting 0.1369882969 equal 0.137 redefines 0.137 centimeter as the distance between adjacent points in the mathematical universe. There may be a number other than 389, closer to the current acceptable distance 0.1369882969, but considering that all mathematical numbers are a product of primes, then 389, also a prime number, is not a bad guess. But it is just that, a guess that in the universe happens to be close to the calculated 400 CMB in a cubic cm. It may just be serendipity, but the analysis has merit, given Richard Feynman's fine structure constant view:

> There is a most profound and beautiful question associated with the observed coupling constant, e, the amplitude for a real electron to emit or absorb a real photon. It is a simple number that has been experimentally determined to be close to 0.08542455. (My physicist friends won't recognize this number, because they like to remember it as the inverse of its square: about 137.03597 with about an uncertainty of about 2 in the last decimal place. It has been a mystery ever since it was discovered more than fifty years ago, and all good theoretical physicists put this number up on their wall and worry about it.) Immediately you would like to know where this number for a coupling comes from: is it related to pi or perhaps to the base of natural logarithms? Nobody knows. It's one of the greatest damn mysteries of physics: a magic number that comes to us with no understanding by man. You might say the "hand of God" wrote that number, and "we don't know how He pushed his pencil." We know what kind of a dance to do experimentally to measure this number very accurately, but we don't know what kind of dance to do on the computer to make this number come out, without putting it in secretly![2]

This is Richard Feynman admitting that neither he or anyone else in physics knows how 137 comes about!

---

2   Feynman, Richard, QED: *The Strange Theory of Light and Matter*. Princeton University Press, p. 129. (1985)

THE FINE STRUCTURE CONSTANT

Consider the distance between adjacent CMB, consider the CMB homogeneously spread throughout the universe, consider the observation that in a given volume, no matter the makeup of the CMB, the number of CMB must offer some resistance to its motion. Then, consider that the shape of the moving object enters into how much drag the moving object encounters. Finally consider the result that a moving sphere attains this value of 137 for the ratio of a speed in a vacuum and the speed in this mathematical medium, with the distance between points a multiple of 137. All these considerations form a dance that nature can follow, and logically puts 137 in without the secret sleight of hand Feynman mentioned. The idea that 137 is $2\varepsilon_o hc/q^2$ is complex compare to the simple explanation that the universe has a homogeneous distribution of particles, as seen in the CMB distribution.

These particles are spheres having a specific volume, a specific cross-sectional area, and a drag coefficient, which at high speed, c, asymptotically approaches the dimensionless number 0.47. This explains 137 without invoking Planck's constant h, charges q, or permittivity ε_o. However, since both formulations result give roughly the same value of 137, we can conclude that within the number of particles in a given volume, the spherical shape, the high spherical speed, and the spherical drag coefficient lies Planck's constant, charge, and permittivity. The realization that the fine structure constant resides in the density, shape, and movement of the CMB is amazing, but not as amazing as the notion that density, shape, and motion of the CMB may give rise to charge, Planck's constant, and permittivity. The mathematical point—a sphere in motion in a sea of spheres in motion—creates approximately the same number, 137, as the results of physics.

Believe it or not, the moving mathematical point makes the concept of continuity and uniformity concrete and understandable. It does not require any proof that the path a moving point outlines, or the mathematical function, is continuous, because the path is continuous. The constant point speed creates a uniform cosmos, where one inch anywhere is one inch everywhere in the universe. This is not the case for a variable point speed. For instance, if in a given region, the point travels at one inch per minute, while in another region the point travels at 2 inches per minute, it becomes clear that mapping of each curve on the same graph conveys different information, therefore limiting the capacity to understand the mathematical functions. Imagine the confusion resulting from a point traveling at any speed it wants to in any given time or space. Uniformity then is a necessary condition in expressing the mathematical function; without space uniformity there is no uniform mathematical function, and no uniform universe.

Another way to say it is that, if the universe were not uniform, there wold not be any mathematical laws.

Although the number of points is infinite, a single point must have space to move. If the point cannot move, then there is no mathematical function. There is a uniform equally spaced number of mathematical point per unit volume, and for a uniform mathematical universe, points are equally spaced from each other just as a glass of water has equally spaced water molecules. Since the points are moving, they offer resistance to the motion of each individual point. Think of walking in a crowded subway car from one end to the other, then compare that speed to what it would be in an empty subway car. The larger the number of people standing in the way, the slower your travel speed through the crowd. The same situation holds for the number of points per cubic centimeter—389—in the mathematical universe. The number is arbitrary, but recall that whatever number one chooses, it must hold throughout the universe. Otherwise the universe is not uniform, and points may have different speeds, which destroys the mathematical universe.

The shape of the point determines how fast the point moves. The spherical volume experiences a resistance to its motion differently than a cubic volume does. Scientists have worked out the resistance generated by different shaped objects, and the numbers they have determined are accurate.

When these factors are taken into account for a moving sphere, they generate the number 137, which mirrors reality, because 137 exist throughout the universe as we know it. It permeates the mathematical universe as well.

Similarly, the number 389 is important, as is the spherical shape, and the fact that the constant speed in a vacuum determines the speed the point travels throughout the entire universe. The number 389 determines the constant uniform distance the points are from each other, and one has to ask: Is it just serendipity that this constant distance is an integer multiple of 137? Certainly, there can be other mathematical universes based on different spherical point concentrations but would they have 137, as the ratio of speed in a vacuum to the speed in the medium? The answer is clearly no.

What is evident is that the number of mathematical point per unit volume is extremely important, as 389 determines the constant universal distance between points in the mathematical universe.

The important aspect to gain from this section is that a moving point creates a continuous mathematical function irrespective of the path the point outlines.

Since the mathematical function exists from minus to positive infinity, the point outlining this mathematical function can never be created or destroyed. If the point were to be destroyed, then the function cannot be defined in the infinite interval; and if the point is created, then the function did not exist prior to the point creation. Thus, no new points can be created, no points can disappear. Because the point cannot be created or destroyed, the infinite number of mathematical points is a mathematical constant.

As an aside, the human implications are enormous: If our make-up is based on this point, we are eternal, and when we die, a part of us will live again.

MOTION, TIME AND, SPACE

Within the mathematical point, a sphere with radius r_o, constant speed v_o, zero charge, zero mass, and moving in the direction i_x, the relationship

$$v_o i_x = X i_x / t$$

indicates that in the sphere, all vectors in the i_x direction are moving with the constant speed v_o. In other words, time t is proportional to X. Since X for the sphere varies, then along X, time must also vary. The time for any moving object varies with the vector length along the motion direction, therefore any time measurement external to the moving object is meaningless, because time may vary in the direction of motion for each vector length in the direction of motion. This mathematical universe is creating a different time concept, which is similar to the difference in time between a traveling and the stationary twin.

Just about every time measurement in physics of a moving object pertains to its external aspects and tells nothing about time inside it. This limits our understanding of an important dimension. Consider as an example a pyramid with a square base b moving along one of its base sides with a velocity b/t at the base. Halfway up the pyramid's apex, the length parallel to the base is $b/2$, and this parallel length must also be moving at velocity b/t. Time for this parallel length must be $t/2$. At the apex, where b is close to zero, t must be close to zero, and here $0/0$ equals b/t. In the direction of motion, varying vector lengths cause varying times. Because time is proportional to vector length, motion alters time.

Another example is the velocity measurements along a right triangle. An object moving at 5 miles per hour along the hypotenuse of the 3, 4, 5 right triangle, is moving with a speed of 3 miles per hour along the 3 side, and at 4 miles per hour along the 4 side. However, the true speed of the object is 5 miles per hour, and along the 3 side, time is not the same as time along the hypotenuse, because along the 3 side, time is $(3/5)t$, and along the 4 side, time is $(4/5)t$; where t is the time along the 5 side. In the mathematical universe, time is different from the time in the Euclidean universe. Keeping time constant for a moving object, may give the wrong speed for lines parallel to the velocity. The solution to many a calculus problems comes up with the wrong answer!

According to this analysis, mv has all of the m lines along the direction of motion **v** moving with speed v. Although mv appears to have two independent variables m and **v**, analysis shows that **v** are all vectors in, and on m. Because the moving object may have different vector length along its line of motion, it may be subject to different times.

The implications are as astonishing as they are profound.

For example, a human body in motion is not a uniform entity. It has different times. Just the way someone sits or stands in a moving object makes a difference regarding the way time operates inside the human body. This concept of motion is fundamentally different from the current idea which posits that objects and motion are independent, and that each can exist without the other. But it turns out that the motion of a body is dependent on the body's volume.

This concept of time may be new to most readers who are only familiar with Euclidean geometry, in which the point has no dimensions and, therefore, nothing inside. But once you posit that a point has volume and explore its internal aspects, you make the surprising discovery that time and distance are proportional to each other there. Since anything that moves can be a point, this internal environment also has the characteristic that time and length along the direction of motion are proportional. They are not independent variables.

Although the notion that time and length are not independent may come as a shock, something resembling this discovery has been discussed in external length contraction, time dilation, and Lorentz transformation But it has never been applied to the inside of an object. Physics and mathematics reduce the object and its mass to a point and ignore the infinite number of internal vector interactions.

VECTOR INTERACTIONS

The mathematical point $(4/3)\pi r_o{}^2(v_o i_x)$ is a vector. Within, and on the sphere, the average vector is $(Xi_x + Y i_y + Zi_z)/2$. These two vectors are simultaneously everywhere in, and on the sphere, and interact to form cross products, dot products, derivatives, curls and divergences. The cross product of these two vectors,

$$(4/3)\pi r_o{}^2(v_o i_x) \times (Xi_x + Y i_y + Zi_z)/2 = (2/3)\pi r_o{}^2 v_o(-Zi_y + Y i_z),$$

is an incredible transformative result. The vector cross product maintains the speed v_o along i_x, and distributes the speed v_o to every vector orthogonal to i_x. The revelation is that this rotational speed comes from the object motion, but more surprising is that every location on the lengths orthogonal to the motion has the speed v_o orthogonal to both, the motion, and length orthogonal to the motion. This is an auto force, and exists no matter the prevailing thought. Although called a pseudo force, the force is real because the force acts on the sphere. In the mathematical universe, there is torque that causes the mathematical point to spin about the axis of motion. All motion must have this torque, which is present in all moving objects. In the current universe this torque "does not exist," and when found, quickly dismissed as "the pseudo force," ignoring the real consequence of torque on the universe.

The moving point, $(4/3)\pi r_o{}^2(v_o i_x)$ is just a sphere moving at the constant speed v_o, and since the point is real, the point must follow Newton's first law of motion:

An object at rest stays at rest, and an object in motion stays in motion, with the same speed, and in the same direction, unless acted upon by an unbalanced force.

The first law states that there are no forces acting on the constant speed body. But there is a torque $(2/3)\pi r^2_o v_o(-Zi_y + Yi_z)$ acting on the constant speed sphere, which is a force. This modifies Newton's first law of motion in such a way that it not only alters physics, but changes the universe. Because the area πr^2_o is a constant, the mathematical point is $(4/3)v_o$, and the torque $\tau = (2/3)\pi r^2_o v_o(-Zi_y + Yi_z)$ becomes $\tau/\pi r_o^2 = B$.

For most lay persons, the impact of torque on the universe may be incomprehensible at first because it redefines physics as we know it. So, let's examine how we reached this conclusion. Most people do not know the rules of vector multiplication, but I can assure readers that the rules were strictly followed (see the Appendix for the rules governing vector multiplication). Torque radically alters physical science and the universe. No doubt, skeptics will challenge this result, but do not let them influence you.

You may not understand the mathematical answer, but to get to the meaning of the result, just bend a paper clip 90 degrees and place it on a piece of paper so that one arm is up and other arm points to your right. Then, take your finger and apply a force to the top of the upright post along the direction of the right post and you will see the torque. You can also see it if you apply your finger to the right post along the downward direction. Viewed in this way, the torque is real, and it is self-evident that whenever a volume moves, a torque acts on the moving volume.

To see this torque in action on its own, take a glass of water, gently pour it, and watch the water rotate about the pouring direction. All motion has this torque.

Torque is a rotating force acting on a moving object. It is present in every moving object. Its existence is a revelation of the highest importance. The presence of torque is a universal law that has no exception. Every moving volume has a torque.

TORQUE

Torque may, or may not cause the sphere to spin, but torque is always present. It changes the direction orthogonal to the motion. The constant speed v_o creates the constant change in direction, f_o, which creates the constant acceleration,

$$a = v_o f_o,$$

where f_o along with v_o are constant throughout the entire sphere. The force acting in the direction of the acceleration is constant throughout the sphere, otherwise the sphere would subject to different accelerations acting on it.

Since $F = ma$, this mass-less point has, all of a sudden, mass. Incredible as it may seem, motion produces mass. Mass is not an independent variable of nature, but dependent on the interaction of the vectors inside an object, as well as on the object's space and time as it exists in a volume in motion. This concept also posits that mass independent of any experimental result or interpretation. Einstein asserted that speed creates mass, but his equations are different. In this case, the equation posits mass where there is none, whereas Einstein created mass from a pre-existing mass m_o that required speed to increase its rest mass.

Fundamentally, mass is the interaction of vectors in and on a moving object, and these vector interactions create mass. In this mathematical universe, such is the power of this torque. Denying its existence, and calling it a pseudo force because it does not adhere to a symmetry principle, takes the quest for the nature of mass along a fruitless path. Answers derived from that approach are wrong. The longer convention stays believing that the torque is pseudo, the longer convention shuns these incredible beautiful revelations.

In this mathematical universe, as well as the current universe, the equations show that spinning objects creates forces and accelerations—let's call them "auto torque"—and therefore, spinning objects create mass.

The conservation of mass principle is no longer true. It is generally accepted that speed creates mass; the greater the speed, the greater the mass. This is not a new concept, as Einstein's relative mass equation shows, $m = m_o / \sqrt{1 - v^2/c^2}$. But because it assumes a nonzero rest mass, it is unable to discover the genesis of mass.

THEOREM 6. *All objects in motion produce an auto torque.*

In the universe, torque is present on, and in every moving object. Therefore, any object or particle in motion has torque—be it a photon, electron, neutron, proton, atom, compound, DNA, RNA, prion, virus, bacteria, cell, plant, animal, human being, bicycle, automobile, boat, airplane, planet, moon, star or galaxy, and even the universe.

MASS

THEOREM 7. *Object motion produces mass in the object in motion.*

Every moving particle produces a mass unto itself, therefore the photon, as a moving particle, has a mass. The photon's mass is not debatable and exists irrespective of the mass measurement results. Imagining a moving particle without mass is science fiction. Understanding that the photon has a spin, the photon's spin is proof that the photon has mass. No argument can be made to deny photon mass, because it leads to $F \neq ma$, and it leads to a change in direction is not an acceleration. Reality has been, is, and will be that the photon had, has and will have mass, no matter what any contrary evidence states. Therefore, with the photon having mass, the equation $m = m_o / \sqrt{1 - v^2/c^2}$ cannot be true. Experiments attempting to "show" the validity of the equation are meaningless. In this mathematical universe, insisting that $m = m_o / \sqrt{1 - v^2/c^2}$ holds, and rejecting anything not in keeping with it, is unacceptable.

For the lay person, it is important to begin to understand the importance of motion and torque, because torque creates mass within a moving object. So important is this mass concept creation that the 2013 Nobel Physics Price was awarded jointly to François Englert and Peter Higgs for a theoretical mass creation concept. In 1964, they proposed a subatomic particle named after the later, the Higgs boson, to explain why particles have mass. (A boson is a particle considered responsible for all physical forces—thus, a photon is a boson.) It is a complicated means to understand an important phenomenon.

But the explanation is actually much simpler. Torque creates mass. Every motion creates mass. There is no need to build equipment to create mass; all that needs to be understood is that motion produces a rotating force that has an acceleration that creates mass. Is there anything simpler than that? No matter what anyone tells you—that this is not mass—just ask, "Is **F** = m**a**?" The answer has to be yes. Is a force acting on a spinning object? The answer has to be yes. Is a change in direction an acceleration? The answer has to be yes. In that case, m = **F/a**. That is the power of logical thinking. Does it need an experiment to be verified? The answer is no.

If an experiment is performed that shows no mass creation, look for an experimental flaw. All experiments are concrete and must use real objects. But experiments also are interpreted in view of accepted theories and facts, and when those theories and facts are incorrect, the interpretation is flawed. However, interpretation based on logic is never flawed unless an axiom is false. That is the problem with Euclidian geometry, which assumes a point "that which has no part." But no such phenomenon exists in real life. The mathematics following Euclid's point is abstract and not applicable to reality. It leads to all sorts of wrong conclusions and interpretations. For instance, the answer to the question, "Can two lines cross and remain in the same plane in reality?" is No. But in Euclid's universe they can, which means that Euclidian geometry cannot exist in reality or accurately describe it.

RESISTANCE

If the torque is strong enough, the moving object spins. Therefore, mass and acceleration are present in every moving object. The sphere is $(2/3)v_o(-Zi_y + Yi_z)$ having already encountered the density, cross sectional area, and the spherical drag coefficient, also encounters the speed squared resistance. The product of these resistances is ε_o having dimension T^2/L^4; where T is time and L is length. When the object has a torque, the object encounters a resistance to the rotation; the rotational resistance is μ_o with dimension L^2. The vector

$$\mathbf{B} = (2/3)\varepsilon_o\mu_o v_o(-Z i_y + Y i_z),$$

has dimension time, T. Nothing external to the mathematical point creates these vectors; motion and volume create an infinite number of vector interactions in, and on the sphere. The moving sphere is a universe where an infinite number of vector interactions occur, and their effects in, and on the sphere create some universal laws. For example, there is the rotational vector $v_o(-i_y + i_z)$ and there is the vector $\mathbf{B}_1 = (1/3)\varepsilon_o v_o X(i_y + i_z)$.

ELECTRIC VECTOR

The vector $\mathbf{B}$ exists in the mathematical point as does the vector $v_o i_x$, therefore the cross product

$$v_o i_x \times \mathbf{B} = (2/3)\varepsilon_o\mu_o v_o^2(-Y i_y - Z i_z),$$

or

$$\mathbf{E}_1 = (2/3)\varepsilon_o\mu_o v_o^2(-Y i_y - Z i_z),$$

and along with

$$v_o(-i_y + i_z) \times (1/3)\varepsilon_o\mu_o v_o X(i_y + i_z),$$

creates a total vector

$$\mathbf{E} = (2/3)\varepsilon_o\mu_o v_o^2(-X i_x - Y i_y - Z i_z),$$

normal to the spherical surface with dimension length, L. This vector $\mathbf{E}$ then defines a charge

$$q = 4\pi\varepsilon_o r_o^2 \mathbf{E}$$

with dimension T^2/L. The creation of a charge from a volume in motion, with the volume having an initial zero charge is breathtaking and awe inspiring. A sphere in motion creates both mass and charge, the two fundamental constituents of matter. In fact, any three dimensional object in motion creates mass and charge.

For the lay reader, what is important to note about the outcome of these mathematical operations is the creation of an electric vector, which then creates a charge. Anything that moves generates a charge, and anything in motion is a mass.

If this essay ended here, it would already have revealed enormous changes, as it derives both mass and charge from volume and motion, or space and time. There is nothing simpler than a volume in motion creating magnetic and electric vectors, and creating the two fundamental constituents of matter charge and mass. This initial discussion establishes a rational, realistic and logical view of the universe, and clearly shows Einstein's genius in his attempt to reduce the universe to space and time, and nothing else.

CHARGE

THEOREM 8. *Objects in motion create charge.*

Every moving object is a charge, therefore the photon is a charge. The motion of an object produces the vector **E**, and this vector produces the charge. Since the charge is moving, it is a current. Thus, a moving charge creates a magnet. The moving mathematical point generates electromagnetism.

Corollary 2. *Based on the cosmic microwave background radiation, the universe is an electromagnetic sea.*

The prevailing electromagnetic concepts of the universe are different from a universe that is a sea of electric charges. Space is not devoid of an electric vector, but rather space is filled with charges. The fact that science calls this space as an electric environment where the electric vector is zero, is just another example of applying a wrong label, which prevents true knowledge from emerging from its eternal prison. The reality is that the universe is a sea of moving charges. Any experimental interpretation stating that the universe is not a sea of charges is plain wrong and incompatible with the cosmic microwave background radiation.

The CMB are electric charges in motion, thus creating infinite energy throughout the universe. In this regard Nikola Tesla was correct. In the universe, infinite energy means infinite energy here on Earth. More importantly, the fact that the Earth spins means that perpetual energy exists. The fact that moons, planets and stars orbit, and that electrons spin and orbit, means that perpetual motion exists. Again, experimental and theoretical findings contradicting perpetual motion, are incompatible with reality, because they deny the Earth's spin, the Earth's orbit, and so on. Similarly, electrons orbiting a nucleus and creating stable atoms is another example of perpetual energy.

For the lay reader, what this means is that beyond the mathematical universe, and into the universe we live in, the cosmic microwave background radiation (CMB) consists of photons. To the extent that they are particles having volume, when they are in motion, they must have a spin, mass, and a charge. These findings contradict prevailing physics dogma. But anything that moves has a charge and mass, as the mathematically manipulations logically prove, and any experimental interpretive results that disprove the photons' mass and charge are flawed and false.

THE AETHER

Because the mathematical universe has an infinite number of spheres each having radius r_0 and constant speed v_0, the mathematical universe has an infinite number of equal magnitude repelling charges. The result is the creation of an infinite dynamic lattice, which operates as follows: If a charge is not at the distance of 0.137 cm from its six adjacent neighbors, repulsion forces the charge to be at that appropriate distance. In the mathematical universe and in the CMB, the repulsive force causes homogeneous distribution. No longer is it tenable to treat the photons as having no mass and no charge, but rather photons form the aether. Another argument showing the aether's existence is: Assume there is no aether, then the mathematical point is $(4/3)ci_x$ because there is nothing to resist the motion. Therefore,

$$\mathbf{B} = (4/3)c(-Zi_y + Yi_z)/2,$$

and taking the curl of this expression results

$$\nabla \times \mathbf{B} = -(4/3)ci_x,$$

which is equal and opposite to $(4/3)ci_x$, hence no motion.

THEOREM 9. *Motion is only possible in an aether, and impossible in a vacuum.*

All the experiments done with electrons and photons in a "vacuum" are incorrect because particle movement cannot occur in a vacuum. It is impossible to create a vacuum in the universe, because the CMB exists everywhere in the universe. The experiment where a photon "becomes a wave" does not take the CMB into account. With particle movement impossible in a vacuum, the CMB creates the wave just as the submarine creates water waves. The photon, or the electron is always in the CMB aether. No particle escapes the CMB aether. Because motion in a vacuum is impossible, the photon is moving in the CMB, which creates the post apertures waves after photons are projected through pinholes.

To conclude that the photon not only becomes a wave, but to become the exact number of waves as there are apertures, is a magical concept, arising from the false belief that the particle is moving in a vacuum that cannot exist. In pre and post aperture , the omnipresent CMB creates the post aperture wave. If there is motion, then there must be an aether. The universe has an aether because all objects in the universe are moving. With mass and charge, the mathematical point and photon offer resistance to all moving objects.

For the lay reader, the argument about the existence of aether has been going back and forth since science started. But in 1887 the Michelson Morley experiment showed no aether around the Earth, although some prominent scientist privately argued that there must be an aether. Again, this mathematical universe comes down on their side: Without some form of resistance, i.e an aether, motion becomes impossible. It would be like trying to walk on a sheet of ice and failing to make any forward progress.

ELECTRON, NEUTRON, AND PROTON

The cross product of the vectors **v**, and **E**

$$B = v \times E,$$

and the cross product of the vectors **v** and **B**

$$E_1 = -v \times B/c^2,$$

which becomes

$$E_1 = -(v^2/c^2)E,$$

therefore

$$E_t = E + E_1,$$

or

$$E_t = E[1 - (v^2/c^2)]$$

where the sum of the electric vector **E** and the magnetic vector **B** create the vector E_t. The equation $E_t = E + E_1$ resembles Lorentz force equation

$$F = q(E + v \times B).$$

The equation $E_t = E[1 - (v^2/c^2)]$ has the following interpretation:

1. For $v < c$, E_t has the same **E** polarity.
2. For $v = c$, E_t is zero irrespective of the E value, or polarity.
3. For $v > c$, E_t has the opposite **E** polarity.

Since the charge q depends on the polarity of **E**, which is negative for the mathematical point and the photon, then q has the negative polarity when $v < c$. When $v = c$, then the particle has no polarity, and when $v > c$, the particle has positive polarity.

THEOREM 10. *The ratio of the speed of the particle to the speed of light in a vacuum determines the particle's polarity.*

With no need to posit the existence of three different particles such as electron, neutron and proton, only one particle is sufficient, and particle speed determines its polarity. In equation $E_t = E[1 - (v^2/c^2)]$, v controls charge polarity magnitude. One moving sphere creates any one of three fundamental particles electron, neutron or proton, but the speed of light, c, turns off electricity and magnetism. Traveling at the speed c, the object looses its electric vector. Orbiting with a speed c, the electron has no attractive force from the positive nucleus. Orbiting a central location with speed c, the positive charge nucleus and the negative charge electron loose their electric vectors. Therefore, the particle speed c is the necessary condition for atom creation.

For the lay reader, anything traveling at the speed of light has no electric vector, and the absence of the electric vector does not allow the particle to interact with any electric vector, but if an atom is made up of positive and negative charges, then the positive and negative electric vectors must be zero, or travel at the speed of light. One of the reasons for inventing quantum mechanics is that there is no way of justifying and explaining the prevention of the orbiting electron crashing into the positive nucleus. The atom is electrically neutral, hence both positive and negative electric vectors must be zero, which translates to the electric vectors traveling perforce at the speed of light. Quantum mechanics came up with forbidden regions in the atom.

The difficulty in having the electron travel at the speed of light is $m = m_o / \sqrt{1 - v^2/c^2}$, or if the electron travels at the speed of light, then the electron mass would be infinite. Given this constraint, it is impossible to formulate an atom with any semblance of reality. With the equation $m = m_o / \sqrt{1 - v^2/c^2}$ being false, any derivation from it is also false.

MAXWELL'S ELECTROMAGNETIC LAWS

With respect to Maxwell's electromagnetic laws, the mathematical point, and the photon satisfy

$$\nabla \times \mathbf{E} = -\partial \mathbf{B}/\partial t,$$

which is $\mathbf{E} = -\mathbf{v} \times \mathbf{B}$, and

$$\nabla \times \mathbf{B} = (1/c^2)\partial \mathbf{E}/\partial t,$$

which is $\mathbf{E} = (1/c^2)\mathbf{v} \times \mathbf{B}$. The divergence of $\mathbf{E}$ is

$$\nabla \cdot \mathbf{E} = (4/3)v_o$$

which is the mathematical point.

The divergence of **B** is

$$\nabla \cdot \mathbf{B} = 0.$$

Understanding that all these derivations are not the results of countless electromagnetic experiments, the vector analysis logic derives these results. However understanding the significance of these results would be extremely difficult if the experiments of giants such as Coulomb, Faraday, Ampere, Maxwell and Heaviside had not taken place. Just as Euclidian plane geometry logically proves that the sum of the angles in a triangle is 180 degrees, this logical analysis proves that a moving object engenders Maxwell's electromagnetic laws. A sphere in motion is able to derive Maxwell's electromagnetic laws, which are considered one of mankind's greatest scientific achievements. In fact, electromagnetic force is one of the four fundamental forces in the universe. Maxwell's impressive achievement is linking the speed of light to the propagating speed of the electromagnetic wave. But in our mathematical universe, it is the speed of the mathematical point that generates the electric and the orthogonal magnetic vectors on the mathematical point. More significantly, these two orthogonal vectors are present on anything that moves. The photon traveling at the speed of light carries the electric and orthogonal magnetic field at the speed of light. An almost logical conclusion, but without the knowledge that the photon is a volume and without the realization that V/t exists. Maxwell's linking the speed of the electromagnetic wave to the speed of the photon is an incredible insight.

For the lay reader, these two orthogonal vectors, electricity and magnetism, are very important, because looking at the inside of the mathematical point derives the magnetic and electric vectors without any knowledge of their existence. Such derivations are unimaginable, yet with logical mathematical operators the entire electromagnetic theory unfolds. That is quite a mathematical achievement, encompassing the universe.

UNITS

To explain the universe, mathematics has to assign units to concepts such as force, mass, charge, magnetism, electricity, heat, work, energy and temperature. The basic definition of the mathematical point is a sphere in motion with the speed units L/T, where L is length, and T is time. Acceleration has units of L/T^2. Volume has units L^3, and area has units L^2. A volume in motion has units L^3/T, with resistance equal to the number of mathematical points in a given volume, or nL^3/L^3, where n is an integer without units. The mathematical point also is Av, where A is the sphere's cross sectional area and $\mathbf{v}$ is orthogonal to the surface area A. Experimentally, for high speeds, there is a resistance that is the square of the speed. The shape of the moving object creates a dimensionless resistance C_d. The product is $n(V/V)AC_d v^2$ which has units L^4/T^2 that are the dimension of $1/\varepsilon_o$. The mathematical point rotates about its axis of motion and encounters resistance μ_o with units L^2. The product $\varepsilon_o \mu_o$ has units T^2/L^2.

The units of $\mathbf{B}$ are $L/T\varepsilon_o \mu_o L = (L/T)T^2/(L^2)L = T$. The units of $\mathbf{E} = \mathbf{v}\mathbf{x}\mathbf{B}$ is L. From $\mathbf{E} = q/\varepsilon_o r^2$, q has units T^2/L. From $\mathbf{F} = q\mathbf{E}$, $\mathbf{F}$ has units T^2. From work is $\mathbf{F} \, \mathbf{L}$, energy has units $T^2 L$. From $\mathbf{F} = ma$, mass, m, has units T^4/L. Since energy is kT/λ where T is absolute temperature, and k is Boltzmann's constant, the units of this is $T^2 L$. It turns out that k has units T^2, therefore temperature, T, has units L^2. From energy hf, h has units $T^3 L$. The energy expression mc^2 has the energy units $T^2 L$. The mathematical point's motion creates heat.

In this mathematical universe, every mathematical expression has units of space and time, therefore every physical constant, and every physical law must be expressed in units of space and time. No concept depends on mass, charge or temperature, as each one of these three concepts are ultimately defined in terms of space and time.

There is no gravitational concept that depends on mass, there is no thermodynamic laws that depends on temperature, and there is no electromagnetic principle that depends on charge. Other than the photon, quantum mechanics does not exist except in the space and time universe, which is different from the accepted non mathematical universe.

For the lay reader, dealing with reality demands a system of units. For instance, drinking eight ounces of milk is different from drinking a gallon of milk; walking eight miles is different from walking eight meters. Units quantify reality. The basic unit of the mathematical point is volume over time, but because the mathematical point only has these units, and is the only object existing, then every derivation from the mathematical point must have space and time. There is nothing else. It is a logical conclusion that there is no mathematical point derivative having units outside these two parameters. The mathematical universe is just space and time.

SPIN AND HEAT

The sphere spins with constant speed v_o, and has a constant force **F**, along the direction of spin. Since everywhere in and on the sphere the speed along the direction of spin has to be v_o, the time has to vary in accordance with the circumference along the spin direction. Spin creates heat. The product of the constant force along the spin direction, and the circumference along the spin direction is work, or heat; therefore, spin produces more work, or heat along the equator than along the poles. Examining the spinning objects such as the Earth, planets, Sun and stars, there is maximal heat production at the equator, and minimal heat production at the poles.

THEOREM 11. *Spin generates heat.*

One implication is that heat differential between the Earth's equator and its poles may have to do more with the spin rather than the proximity of the equator to the Sun. The CMB are spinning photons and create heat that may be part of the universe being at 2.75 Kelvin (-250.65 Celsius or -455.17 Fahrenheit). Given the omnipresent photon spin, perhaps the 2.75 K may not just be due to the "expanding universe".

In summary, a constant speed sphere outlines a continuous uniform mathematical function, and motion causes the sphere to rotate about the axis of motion. The rotating sphere is a magnetic vector leading to an electrical vector. Rotation creates mass and charge in all moving objects. From these moving rotating mathematical points, we derive Maxwell's electromagnetic laws. Taking the prime number of 389 as the number of mathematical points in a cubic cm leads to the fine structure constant and the concept that motion requires resistance—or an aether—as well as the notion that spin creates heat. The density of photon, 400 per cubic cm, resembles the arbitrary mathematical point density of 389 per cubic cm. Therefore, the photon helps keep the universe warm.

An object orbiting another object creates heat because the force causing the object to orbit acts along the orbital path. The perpetual orbit creates perpetual heat. Similarly, with perpetual spin, the Sun's heat is perpetual. No need to worry about the Sun loosing heat. The photon spin creates heat, a consequence of its linear motion, which according to Newton's first law is perpetual, therefore heat is perpetual, and has nothing to do with the Big Bang. The reality is that spin produces heat. This is a new concept, even though spin has been present since motion started. An object having a constant speed produces heat, which does not appear in any of the equations, except for mv^2, but among physicists, heat is not considered along the same line as kinetic energy. Although the volume in motion produces spin, magnetic and electric vectors, mass, charge, force, and heat, when it comes to reality something is still missing.

Although linear motion, spin and heat creation occur, there is still no mechanism for other creations. If it is true that the mathematical point is a charge, which has a constant speed, how are neutrons and protons created in light of the mathematical point having a constant linear speed v_o? How is the mathematical point going to go faster than v_o?

The mechanism that makes this possible may follow this broad outline:

With $x = (4/3)\pi r_o^2 v_o i_x r_e \cos nf_e t$, the sphere $(4/3)\pi r_o^2 v_o i_x$ moves along the x-axis with a constant speed v_o, a constant frequency nf_e, a range of t from $\pm\infty$, and a range of x from $\pm r_e$.

With $y = (4/3)\pi r_o^2 v_o i_y r_e \sin nf_e t$, the sphere $(4/3)\pi r_o^2 v_o i_y$ moves along the y-axis with a constant speed v_o, a constant frequency nf_e, a range of t from $\pm\infty$, and a range of y from $\pm r_e$.

The linear movement along the x-axis causes the sphere $(4/3)\pi r_o^2 v_o i_x$ to spin about the x-axis, and this spin causes the sphere $(4/3)\pi r_o^2 v_o i_y$, through action at a distance, to orbit with a radius r_e in the x-z plane. The linear movement along the y-axis causes the sphere $(4/3)\pi r_o^2 v_o i_y$ to spin about the y-axis, and this spin causes the sphere $(4/3)\pi r_o^2 v_o i_{x'}$ through action at a distance, to orbit with a radius r_e in the y-z plane. The linear motion creates the rotation about the axis of linear motion. This represents not a new law, just the same result of any moving object. There are two circles with each point orbits its respective linear motion axis with an orbital radius r_e and orbital frequency nf_e.

Destruction of the orbital circles results in two mathematical points each traveling at speed v_o and orthogonal to each other.

In a sphere of radius r_e with a spinning speed $2\pi r_e nf_e$, the integer n can be very close to infinity, therefore the rotational speed of the two mathematical points creating the sphere can be very close to infinity. The linear speed of the mathematical sphere is always v_o, but the motion of the two orthogonal mathematical points results in two circles which each create a rotating sphere with a rotating speed very close to infinity.

The only motions the point has are the constant linear movement and spin. No law exists in the mathematical universe that would requires the mathematical point to outline any mathematical function other than the straight line. However, because of its linear motion, the mathematical point spins, creating a circle that gives rise to sines and cosines. To the extent that space is influenced by this motion, the aether may also be spinning. Given the existence of the aether, the mathematical point's spin causes the aether to rotate.

The galaxies demonstrate such circular motion. They appear to have all their stars in a plane, orbiting a black hole. Each individual star may have a group of planets revolving around it, and each planet may have a number of moons acting in a similar manner. A galaxy is a larger version of the solar system.

Utilizing spin, the mathematical sphere cannot outline a sinusoidal function, however, and de Broigle's pilot wave theory (or Bohmian mechanics) does not hold. But a collision between two mathematical spheres causes the spheres to vibrate, and these vibrations outline sinusoidal paths, which can be expressed in sinusoidal mathematical functions.

Therefore, the only motions in this universe are linear motion, rotational motion, and sinusoidal motion. This opens up a wealth of possibilities, from Schrodinger's wave equation to string theory.

In plane geometry, nothing happens until the lines intersect. Similarly, in the mathematical universe, nothing happens until the points collide. The collision of point A, $(4/3)\pi r_o^2 v_o \mathbf{i}_x$, with point B, $(4/3)\pi r_o^2 v_o \mathbf{i}_y$, yields $(4/3)\pi r_o^2 v_o \mathbf{i}_y$ for point A, and $(4/3)\pi r_o^2 v_o \mathbf{i}_x$ for point B. The two points exchange their vector and spin directions, and each point vibrates along its axis of motion.

THEOREM 12. *Collision between two identical spheres causes an exchange of vectors and spin.*

The sphere is the only shape that leads to Theorem 12 because its cross sectional area is the same no matter its travel direction. Collision of more than two points destroys the mathematics of the colliding points. Take the collision of the three mathematical point A, $(4/3)\pi r_o^2 v_o \mathbf{i}_x$, point B, $(4/3)\pi r_o^2 v_o \mathbf{i}_y$, and point C, $(4/3)\pi r_o^2 v_o \mathbf{i}_z$, which results in the ambiguous A locations $(4/3)\pi r_o^2 v_o \mathbf{i}_y$ and $(4/3)\pi r_o^2 v_o \mathbf{i}_z$, or anywhere in between the 90 degree angle in the y-z plane. Similar ambiguity results for points B and C. The interaction of three or more bodies has no unique solution.

TWO POINT COLLISION

THEOREM 13. *At any given time, no more than two mathematical spheres collide.*

Since there can be no collisions of three or more spheres, the mathematical universe is planar. The evidence appears in the galaxies, solar systems, planetary orbiting moons, and asteroid with orbiting moons.

The mathematical universe has many different point orientations. As an example, consider just 4 different point orientations. One type exists at positive infinity $(4/3)\pi r_o{}^2 v_o i_x$ traveling towards negative infinity, while another exists at negative infinity $(4/3)\pi r_o^2 v_o i_x$ traveling towards positive infinity. Yet another exists at negative infinity $(4/3)\pi r_o^2 v_o i_y$ traveling towards positive infinity, and the final type exists at positive infinity $(4/3)\pi r_o^2 v_o i_y$ traveling towards negative infinity. When $(4/3)\pi r_o^2 v_o i_x$ collides with $(4/3)\pi r_o^2 v_o i_y$ at negative infinity, they create a staircase going from minus infinity to positive infinity. Rotating the staircase 45 degrees creates a right triangular wave having two equal sides of 0.137 cm and a base that is $\sqrt{2}(0.137)$cm. These waves are the paths the CMB photons take. To ascertain if this is even close to CMB reality, consider that the wavelength where the CMB has maximal radiation is about 2 mm, which is very close to the triangular wavelength of 1.93747258 mm. But one should note that while the two figures approximate each other, it is no proof that the wave is as described.

Traveling along i_x, the pre-collision mathematical point becomes the post collision mathematical point i_y. This transition occurs over a finite distance and finite period of time. Dividing the distance between adjacent points, 1.37 mm, by the integer v_o yields a distance λ_e. The time between collision is $t=1.37\times10^{-3}/v_o$. While the two points are in contact, one must go through a deceleration to lose its speed along i_x, and an acceleration to gain the speed v_o along i_y. Because there is a change in direction, a force must be present for a distance λ_e. To maintain uniformity in the mathematical universe, the point travels at speed v_o yielding the time λ_e/v_o to cover the distance λ_e. The force is present for the distance λ_e, and the product of force and distance is work, or heat. The bodies in the collision gain the resulting heat. The colliding bodies gain the resulting heat.

This leads to:

THEOREM 14. *Mathematical point collisions produce heat. CMB photon collisions produce heat.*

VIBRATION

In the process of colliding, two spheres vibrate, and their vibrations produce heat. Afterward, each sphere continues to vibrate until the next collision, a distance of 1.37 mm away. Think of this process as two cymbals clashing with a time interval of $1.37\times10^{-3}/v_o$. We can call this the beat of the universe, or the metronome of the universe. The sphere vibrations follow a harmonic series pattern. Initially the sphere travels the distance λ_e, initiating the harmonic series $1 + 1/2 + 1/3 + 1/4 + \cdots + 1/n$. This it is not a convergent series. Divergence does not happen in nature.

The convergent sequence begins with the vibrating sphere traveling the distance increment λ_e. From there, the next increment it moves is the product of 1, the first term of the harmonic series and the last distance the sphere traveled, λ_e, yielding $(1)\lambda_e$. Therefore, the total distance the sphere has traveled thus far is λ_e + $(1)\lambda_e$. The product of the next harmonic term 1/2 and the last increment the sphere traveled, $(1)(\lambda_e)$, yields the next increment,$(1/2)$ $(1)(\lambda_e)$ and results in a total distance traveled of $\lambda_e(1+1+1/(2(1)))$. The next distance increment is the product of (1/3) and the last increment $(1/2)(1)(\lambda_e)$ yielding a new increment, $(1/3)(1/2)(1)(\lambda_e)$ and resulting in the new covered distance $\lambda_e(1+1/1!+1/2!+1/3!)$, and so on. For the nth harmonic term the sum is $\lambda_e(1 + 1/1! + 1/2! + 1/3! + \cdots + 1/n!)$ and the total distance converges to $\lambda_e e$; where e is the transcendental natural number $e = 2.718281828459045\cdots$.

This converging process is the genesis of e. Any vibrational process with an initial value of $\lambda_e \neq 0$ creates e. Vibration occurring in this sequential harmonic manner create e, rather than a dictum declaring that it is e because it is the measured value. In this mathematical universe, space and time generate every mathematical function. It is not the mathematical function that creates space and time. To produce e, an object has to be in motion in space and time.

THEOREM 15. *Collision causes vibrations, and the vibrating object traces interacting sinusoidal paths culminating in the natural number e.*

A natural process, vibration generates the harmonic series. Every vibration process does so. In addition, because each harmonic term occurs in the harmonic series sequence—i.e., 1 comes before 1/2, which comes before 1/3 and so on—it also generates e. The sequence terms occur in decreasing time intervals. Without knowing anything about calculus, the mathematical process creates e as the time increment approaches zero. The mathematical point's vibration creates calculus, not the other way around.

To give a concrete visual example of the harmonic sequential process, consider the vibrations of a plucked violin string. The new harmonic term interacts with a nonzero increment, which is the product of the harmonic term, and creates a new increment, which is the point's new travel increment that adds to the total covered distance. Each subsequent point's travel distance converges rapidly towards zero. The sum of all these distances is the natural number e. But, the greater mathematical significance is the point's vibrations outline sinusoidal mathematical functions, and with the product of the harmonic series generates the natural number e. There may be other processes generating e, but the natural two point collision process shows a relationship between the natural number e with mathematical sines and cosines functions. Leonhard Euler's remarkable insight showed the relationship between e, $sin(x)$, and $cos(x)$ through the equation $e^{ix} = \cos x + i \sin x$.

THE TEMPERATURE OF THE UNIVERSE

The measured temperature of the universe is 2.725 K, and is coincidental with the mathematical theoretical number $2.718281828\cdots$, with v_o, the infinite number of collisions happen approximately every $1.37 \times 10^{-3}/v_o$ of a second. These collisions started at the creation of the universe, occur now, and will continue to happen as long as the universe exists. The infinite number of point collisions causes and preserves the heat in the universe. Although the measured temperature is 2.725 K, and the mathematical theoretical temperature is 2.718281828· · · K, there is also heat created by the point spin, which may account for the discrepancy.

Again, space and time create the temperature of the entire universe through the colliding and rotating volumes, yielding a close approximation to the carefully measured temperature of the CMB universe. The idea that the temperature in the universe remains constant runs counter to the Big Bang theory, which contends that temperature decreases due to expansion. But the infinite mathematical universe is not expanding, and no explosion created it.

The mathematical point moves in a back and forth motion along the propagating direction, outlining a longitudinal wave. The mathematical point has the same speed v_o and direction i_x with the point's to and fro motion direction i_x. While the interaction of the harmonic series with the sphere determines the magnitude of the to and fro motion, the distance is not subject to that motion.

The speed v_o is, according to the formula

$$(\lambda_e/n)nf_e \sin 2\pi(nf_e t)i_x,$$

where λ_e is the wavelength, f_e is the frequency, t is time, and n is an integer. It is "understood" that the "wave" has an amplitude v_o, which propagates along the positive x-axis with a speed $(\lambda_e/n)nf_e = v_o$.

The mathematical point actually undergoes this to and fro motion and advances with the constant speed v_o in the propagating direction. The speed of the mathematical point is always v_o no matter the frequency or wavelength. The maximal wavelength is λ_e, and the minimal wavelength is $2\pi r_o = \lambda_e/n$. Along the motion direction, the longitudinally vibrating sphere experiences a force over a total distance $e\lambda_e$. This force only occurs while the sphere changes direction and not when the sphere is moving along a straight line.

The higher the vibrating frequency, the higher the heat production. The amplitude of the sinusoidal is immaterial with respect to heat generation, because frequency generates heat. Certainly, the sinusoidal must have an amplitude, but what generates the heat is directional change frequency. For a given sinusoidal amplitude, v_o, the higher the number of back and forth motion, or frequency, the higher the heat production. This can be demonstrated by taking a wire, and bending up and down; the more frequent the bending, the higher the heat production. It does not matter how far the bending arm is from the bending point, what matters is the bending frequency.

THE CREATION OF THE UNIVERSE

We have already discussed that the concentration of the number of mathematical points in a given volume is 389 points per cubic cm, and that this concentration determines the number of mathematical points in a cm. Depending on the way the cm is defined, the distance between adjacent points is exactly 0.137 cm. All positive integers, except for the integer one, are products of prime numbers, and since the speed v_o equals $1.37 \times 10^{-3}/t$ and must be an integer, then the simplest $v_o = 137P$, where P is a prime number. It is also true that $v_o = 1.37 \times 10^{-3}/T = 137P$, or $TP = 10^{-5}$. Assuming that 299792458 is a close measurement of v_o, then 299792458/137=2188266.177, which is not an integer; but 2188267 is an integer and is a prime. There-fore, v_o = 299792579, and T has the unique value $10^{-5}/2188267$ of a second.

The two prime numbers 137 and 2188267 determine all the constants in this mathematical universe, but given all the mea-surements to get to 299792458, the value 299792579 approxi-mates the measured value. The crucial aspect is that the length and the time are both fixed in the mathematical universe, and the same may hold in the cosmos through the photon.

SOME PHYSICAL CONSTANTS

Let us briefly examine the prevalence of the two prime num-bers in various aspects of physics. Assigning the value of (137) (2188267)ms^{-1} to v_o, enables physical constant determinations. The magnitude of the constants is compared to the magnitude of the corresponding National Institute of Standards and Technol-ogy (NIST) constant.

For electromagnetism, the permeability constant, μ_o is (2/3)L^2 with L=1.37x10^{-3}m, which yields 1.251266667x10^{-6}m², and the NIST value is 4πx10^{-7}, which equals 1.256637061x10^{-6}. With time, T=L/v_o, having the value 1.37x10^{-3}/((137)299792579)= 4.56982626x10^{-12}s, the permittivity constant, ε_o, is (3/2)T^2L^{-4} and has the value 8.892182518x$10^{-12}s^2m^{-4}$, while the NIST value is 8.854187817^{-12}.

In the energy equation E = kT, where k is Boltzmann constant, and T is the absolute temperature in Kelvin, but with T being time in the k expression, k= (2/3)T^2 is 1.3922208803x$10^{-23}s^2$, and the NIST value is 1.38064852x10^{-23}.

For Planck's constant, which deals with the actions of photons at the atomic level, $h=(2/3)T^3$ is $6.362207185 \times 10^{-35} s^3$, and the NIST value is $6.626070040 \times 10^{-34}$, which is off by a factor of about 10.

The spherical constant, 4/3; the number of mathematical points in a given volume, 389 points in one cubic cm; the distance between adjacent points, $L=1.37 \times 10^{-3} m$; the inverse of the fine structure constant: 137; and the prime number 2188267 are the only constants creating the constants μ_o, ε_o, k and h. They open the portals to unifying the mathematical universe. Along with the spherical mathematical point's motion, these constants make their appearances in magnetism, electricity, heat, light, charge and mass.

For the lay reader, these constants are found in various areas of physics. Using the numbers 137 and 2188267 from the mathematical universe allows us to generate them as well. Although the units are not the same, their values are close to each other, lending credence to the mathematical universe. The units not being the same is to be expected because the physical universe, in addition to space and time, also has mass, charge and temperature. The mathematical universe just has space and time, and all constants are derived accordingly.

A POINT IN THE MATHEMATICAL UNIVERSE

Motion in the i_x, and i_y directions limits the mathematical universe, but $\cos \psi i_x$, and the orthogonal $\cos \psi i_y$ rotates the axis, and is more reflective of the mathematical universe. In the solar system, the different planetary axis of rotation, or inclination angle, suggests collisions are between $\cos \psi i_x$, and the orthogonal cos ψi_y, and not just between i_x and i_y. In a given galaxy, all stars orbit the black hole. Since all galaxies have an inclination angle, the black hole has an inclination angle, and the black hole spins. Only spinning objects cause the space to move around it.

Every location of the mathematical point has the spin speed v_o. All space external to the rotating mathematical point must spin about it. In any specific location, the expression for a constant speed rotating sphere, has the acceleration $a = v_o f_o$.

In this mathematical universe, when a black hole is formed, it compresses everything. Thus, it causes the concentration of the mathematical points to increase from 389 per cubic cm with a 1.37 mm distance between adjacent points to a $\lambda_e = 1.37 \times 10^{-3}/v_o$ = $4.56982626 \times 10^{-12}$m distance, which is the minimal distance between adjacent mathematical points. At this minimal distance, the mathematical point interactions create the f_e spin frequency electron. As the distance between mathematical points increases, the interactions result in $2f_e$ spin frequency electron, and so on. Generally, as the distance between mathematical points increases, their interactions create the nf_e spin frequency electron. However, when n is 137, the process creates a neutron. For all values of n greater than 137, the mathematical point interactions result in positive charge particles.

In this mathematical universe, as in the chemical universe, all atoms—and chemical elements and compounds—are electrically neutral, or have a zero electric field. Most scientists, however, think that an electron with a negative charge cancels an equal magnitude positive charge proton. But that is not the case. An electron separated from a proton by a distance would create a dipolar electric field—positive and negative charges separated by a distance—which does not have a zero value everywhere. To attain a zero electric field, all the electrons of an atom must orbit with the speed of light, c, and all the protons must orbit with speed c as well. That is the only way an atom can have an electrical vector equal to zero.

For the lay reader, this electrical neutrality is extremely important. There is a rule that the number of electrons must equal the number of protons, canceling out each other's negative and positive charges. But that idea is erroneous. A negative charge at distance from a positive charge does not produce a zero electric field, but a non zero electric field called a dipole electric field. If atoms are composed of charges, then they are not electrically neutral. If charges are present, then it is impossible to have a zero electric vector. Electrical neutrality is the rational for the rule that an atom must have an equal number of electrons and protons. But because it is impossible to create a zero electric field, the rule may be correct, but not because of electrical neutrality. If this is the case, then traditional chemistry is not correct.

The way around the problem is to have the object travel at the speed of light, c. Then the electric vector is zero. All orbiting particles must have an orbital speed c. The electrons and protons, and their charges, must orbit a location with a speed equal to c.

This is an extremely difficult notion for most scientist to accept. They prefer to believe that two particles of equal magnitude opposite polarity create an electric field that is everywhere zero electric field. But that has not been experimentally verified. In fact, I would argue that it can't be done. To the naysayers, I would suggest, since science is formulated on the basis of experimental results, have them show that a dipole field has a zero electric field, and if they cannot do so, perhaps they will have an easier time accepting some of the consequences of the approach detailed in this essay.

Proton spin frequency determines the electron orbiting speed, and the electron spin frequency determines the proton orbital speed, where f_{es} is the electron spin frequency, f_{ps} is the proton spin frequency, veo. The proton interacting with the f_e electron to create a hydrogen atom is the proton with a spin frequency $f_p = 137^2 f_e$. All electrons are integer multiples of f_e, and all protons creating stable atoms are integer multiples of $137^2 f_e$. The ratio of the spin frequency of the electron, f_e to the spin frequency of the proton, f_e/α^2 is always α^2. The product of the orbital speed of the electron and the orbital speed of the proton must be the speed of light squared:

$$f_{es}/f_{ps} = \alpha^2,$$

and

$$v_{eo} v_{po} = c^2;$$

where v_{eo} is the orbital speed of the electron, and v_{po} is the orbital speed of the proton.

The electron electric vector is

$$E_{et} = E(1 - v^2_{es}/c^2),$$

and the proton electric vector is

$$E_{pt} = E(1 - v^2_{ps}/\alpha^4 c^2),$$

The two electric vector's magnitudes are not equal, but when the orbital speed of the electron is c, then the electric vector is zero. The same holds true for the proton.

The speed creates the electric vector, which creates the charge. Throughout this mathematical universe, speed dictates what occurs. Whenever an object travels at the speed of light, its electric vector is zero. It is speed that makes atoms possible and eliminates some quantum mechanics rules. Atom creation would be impossible without $E_{et} = E(1 - v^2_{es}/c^2)$, which is the equation responsible for all the chemical elements in the periodic table. This equation is nothing more than Lorentz electromagnetic force equation $F = q(E + qvB)$, and has been known for well over a century, but science believing $m = m_o/\sqrt{1 - v^2/c^2}$ is compromising the power and beauty of $F = q(E + qvB)$.

For the lay reader, the equations may be a bit difficult to understand, but the idea is that atoms can only form when the electric charges are traveling at the speed of light. Why should this be important? It turns out that all of chemistry is founded on that one equation.

Although the mathematics may be a bit overwhelming, the equation $E_{et} = E(1 - v^2/c^2)$ is worthwhile remembering and understanding. It describes the creation of an electron when the speed of the object is less than the speed of light, c, and the creation of positive charges when the speed is greater than c. But when the speed of the object is equal to c, not only does it create a neutron, it makes atoms possible. Indeed, it makes every object in the universe possible, including life itself. Everything in the universe depends on the speed of the object. I hope that the reader can begin to feel the significance of time as it appears in the speed and of space as it appears in speed and volume.

It is possible that there are atoms forming between electrons and protons that may have protons that are not integer multiples of $n = 137^2 f_e$. These protons create isotopes. To create a stable atom, such as hydrogen, the electron must orbit the center of the atom at a speed c—the speed of light in a vacuum—otherwise the mathematical point would have a charge. Therefore, at a distance λ_e/α, the distance the electron is from the center of the hydrogen atom, only the f_e/α^2 spin frequency proton causes this electron to orbit at speed c. Similarly, the f_e spin electron causes the $137^2 f_e$ spin frequency proton to orbit at a distance $\lambda_e/137$ and have an orbital speed c.

In this mathematical universe, the electron and the proton must have specific orbital radii. Thus, the electron orbital radius is r_e/α, and is not a probability function. Because the electron's orbital speed is c, there is no electron uncertainty about position and speed. In this mathematical universe, Schrodinger wave equation as a probability function leads to an unrealistic universe. Einstein was correct in positing that the speed of a photon being constant is the necessary condition for a mathematical, and therefore, a deterministic and precise universe.

In this mathematical universe, the electron mass is m_e and through the equations $E = mv^2_o$ and $E = hf_e$, m_e, and f_e are proportional to each other. Since f_e is a quantum, the electron mass is a quantum.

The electron spinning with a frequency $2f_e$ has a mass $2m_e$. The electron with spin frequency nf_e has a mass nm_e.

A proton with a spin frequency f_e/α^2 has a mass m_e/α^2. A proton having twice the mass of this proton is a proton having the frequency $2f_e/\alpha^2$. The proton frequency as well as the proton mass are quanta.

An electron with spin frequency f_e has a maximal orbital frequency f_e. The electron's orbital frequency cannot be greater than the electron's spin frequency, otherwise the mathematical points would be forced out of the electron and lead to electron destruction.

All atoms are charge neutral. Both electron and proton must move with a speed c, the only speed that causes all charges to become zero. The electron or proton having a speed other than c, are charges, which can never be made zero. For orbits, the electron orbital speed must be c, and the proton orbital speed must be c.

In this mathematical universe, Hydrogen is the first element in the periodic table. The next element represents a bit of a problem, however. Rather than Helium, I would argue it is an element Deuterium, consisting of 2 protons and 2 electrons, currently considered an isotope of Hydrogen.

There are various options to consider. The first is the electron with a mass $2m_e$ and a proton with a mass $2m_p$. The other is having two electrons each of mass m_e but having opposite orbiting directions, and two protons each having a mass m_p, but having opposite orbiting directions.

This has been overlooked until now. Einstein said that energy equals mass times the speed of light squared—the famous formula $E = mc^2$. But he also posited that the energy of a photon is governed by its frequency multiplied by Planck's constant. Since the electron orbiting a point at any orbital frequency creates mass, $mv^2 = hf$, it means that frequency and mass are equivalent!

Notice that the significant mass added to the hydrogen atom is $m_p + m_e$. That is what conventional physicists are calling the mass of the neutron, but the mathematical neutron has the mass $m_n = m_e/\alpha$. In this mathematical universe, there is no need to postulate a neutron in the nucleus, which only contains protons.

The next important question concerns the radii of the electron and proton. The radius of the nucleus is said to be about 10^{-15} of a meter. A radius of $r_n = \alpha\lambda_e/2\pi$ equals $5.308835316\times10^{-15}$m. The NIST Bohr radius is $a_o = 5.29177210903\times10^{-11}$m, and in the Hydrogen atom, the electron orbital radius is $r_{eo} = \lambda_e/2\pi\alpha = 9.964153005\times10^{-11}$m. The radius of all protons is the same as the electron. The electron has the relationship $2\pi r_e f_e = v_o$.

All mathematical points move with a linear speed v_o, spin with constant frequency f_e, have constant mass, and have a constant charge. Therefore, the mathematical point cannot be transformed into anything else. How then can there be particles such as electrons, protons, and neutrons? Collision causes point vibrations, which are harmonic vibrations, which are sinusoids that can form circles. What radius an electron has is not all that clear. The current understanding posits a point charge with no spatial extent. Again, the Euclidian concept creates an imaginary point, but it also creates an imaginary energy, as well as an imaginary charge. Therefore, energy requires nothing to produce it; that is to say: Force is not necessary to produce energy, motion is not necessary to produce energy, heat is not necessary to produce energy, electromagnetism is not necessary to produce energy, the strong nuclear force is not necessary to produce energy, the weak nuclear force is not necessary to produce energy, and gravity is not necessary to produce energy.

When the total energy is incorrect, the imaginary energy corrects it, and erases any contradiction in quantum mechanics.

But we must ask ourselves: An atom, the fundamental unit of the universe, has an imaginary electron orbiting a proton? How can that be? Is this is reality? The imaginary electron has no spatial extent, in which case, an infinite number of electrons create an electron with an infinite number of electrons. Think of zero: how many zeros can one fit in zero space? An infinite number. But that bears no relation to reality. It's what happens when something "has no part." Continuity becomes an arbitrary distance between two electrons. But however small one can make that length, there is always a distance smaller than that. Positing no spatial extent is the epitome of abstraction, not that different from "that which has no part," which causes mathematics to create abstractions that have no equivalent in reality. If the electron has no extent, then physics is an abstract subject, chemistry is an abstract subject, biology is an abstract subject, and the universe is an abstract subject. Nothing would exists, but this is not the case. There are atoms. Therefore, the electron has to have a spatial extent, and that part or extent is volume. If quantum mechanics requires that the electron has no extent, then quantum mechanics describes a universe that does not exist!

The electron having mass but no spatial extent results in the two fundamental constituents of matter, mass and charge, being imaginary. Everything in this imaginary universe is possible, including parallel universes in which identical events occur at the same absolute time. That is the stuff of science fiction.

In the real universe, all particles have spatial extent, or volume. The nuclear radius is in the range of 10^{-15}m. The Bohr radius $r_{Bohr} =$ 5.29177210903x10^{-11}m. The mathematical $\lambda_e =$ 4.56982626x10^{-12} results in $r_{Bohr} =$ 9.96415300x10^{-11}m and is the orbital radius of the electron. The relationship between the spherical radius and wavelength is $2\pi r_e = \lambda_e$, therefore, the mathematical electron radius $r_e =$ 7.273104383x10^{-13} has a spatial extent. The mathematical electron radius is the radius for the electron, while the proton and neutron may have different radial extent. The relationship between the spherical radius, r, and the spherical wavelength, λ, is

$$2\pi r = \lambda$$

There are no imaginary particles in the realistic mathematical universe.

PARTICLE CREATION

The mathematical point traveling along the positive x-axis $(4/3)\pi r_o^2 v_o i_x$ collides with the mathematical point $-(4/3)\pi r_o^2 v_o i_x$ traveling along the negative x-axis. In the process, they behave just like two billiard balls; they collide, exchange directions, vibrate and do not disappear. The spherical vibration occurs along the x-axis only, and the vibrating speed is the product of the wavelength and frequency. The vibrating equation is

$$\pm(\lambda_e/n)nf_e \cos(nf_e t)i_x,$$

where n is an integer from one to infinity. Since the vibrating object is a sphere, $\pm 2\pi(r_e/n)nf_e \cos(nf_e t)i_x$ holds, and the total equation is

$$\pm[(4/3)\pi r_o^2 v_o i_x][2\pi(r_e/n)nf_e \cos(nf_e t)i_x].$$

This is a mathematical point moving with velocity $v_o i_x$ for a total distance 1.37×10^{-3}m, when the next collision occurs, causing $-v_o i_x$. During the time that $v_o i_x$ exists, $\pm 2\pi(r_e/n)nf_e \sin(nf_e t)i_y$ exists, and causes all mathematical points to rotate about the positive x-axis.

Similarly along the y-axis, the equation that obtains is

$$\pm[(4/3)\pi r_o^2 v_o i_y][2\pi(r_e/n)nf_e \sin(nf_e t)i_y],$$

and the mathematical point $[(4/3)\pi(r_o^2/n)v_o i_y]$ forms a circle about the x-axis just as $[(4/3)\pi(r_o^2/n)v_o i_x]$ forms a circle about the y-axis.

These two orthogonal circles have a radius r_e, and each point has a circular speed $2\pi r_e nf_e = nv_o$. Each point creates a rotating sphere whose circular interaction gives the appearance of a sphere spinning about a 45 degree axis.

There may also be four circles consisting of two circles rotating in the same direction and two orthogonal circles rotating in the same direction. This yields a sphere rotating at an axis of 45 degrees. However, if two circles each rotate in opposite directions and two orthogonal circles each rotating in opposite directions, the result would be a non-rotating sphere. Rotation in the same direction causes the entire universe to orbit the spinning sphere. Opposite direction rotation does not. Irrespective of spin direction, each sphere has two opposite centrally acting forces. The difference between these two forces is gravity. Non-spinning moons have gravity, but do not cause the universe to orbit them.

These spheres create electrons, protons, neutrons, moons, planets and stars. The interactions of electrons and protons creates atoms.

After the collision, each of the two mathematical point vibrates with the discrete frequencies in the harmonic series, having the highest vibrating frequency just prior to the next collision. The vibration is a sinusoid and, along with the harmonic series, each subsequent term may be of the form

$$(\lambda_e/n) \sin 2\pi nf_e t,$$

where n is an integer from one to a very large number, which is determined by the inverse of the vibrating sphere's radius. This harmonic process is similar to the sequence we discussed earlier, using the example of a plucked string. If the first displacement is 1, the second displacement is 1/2, the third displacement is 1/3, and the nth displacement is 1/n th. The first particle to form is the black hole with a sphere having a spinning speed of $\lambda_e n f_e$, which is much greater than c, and therefore has the highest positive charge in the universe.

As a result, it does not allow any spin of magnitude in the same direction in its region. The huge positive charge attracts the surrounding negative charges, causing them to orbit at a speed greater than c and transforming them into a positive charge particles. As two opposite spin positive charges attract, they abut the center and create a larger spinning sphere, which rotates at a lower frequency than the frequency of the central sphere. The adjacent particles adjoining each other increase the number of particles per cubic cm. In the process, they slow down the newly formed sphere to a slower spin speed than would exist if there were 389 particles per cubic cm. The process continues until the spinning sphere's speed reaches c, the speed of light in a vacuum. With the distribution of mathematical points greater than 389 points per cubic cm, the mathematical points have a speed less than the rotating sphere should have at that radius. The difference in speed creates gravity until the mathematical point concentration becomes 389 and the outward radial force equals the inward radial force of gravity, resulting in a net radial force of zero.

The concentration of particles leads to a distance between particles of λ_e, which is much smaller than 1.37 mm, the distance between mathematical points. At distance λ_e, the only particle that forms post collision is the electron, because the sphere cannot vibrate more than once in that interval before the next collision occurs. At distance $2\lambda_e$, the vibrating sphere can vibrate twice and creates an electron having a frequency $2f_e$. As the distance between adjacent particles increases, more electrons, protons, moons, planets, and stars are created until, finally, when the distance is 1.37 mm, black holes form.

The interaction of electrons and protons creates the elements in the periodic table.

The rule that allows for an electron and a proton to interact, and avoid collision, is as follows: An electron must orbit a location with an orbital speed c, and the proton must do the same. This is the only way that an atom can be electrically neutral or have a zero electric vector.

The electron spin causes the proton orbit, and the proton spin causes the electron orbit.

Although is true that an electron may have a frequency f_e, there are 136 electrons having frequencies ranging from f_e to $136f_e$. But the positive charges can have frequencies from $138f_e$ to nf_e, where n is a very large positive integer.

In this mathematical universe, the atom requires no neutrons. In any atom, there is only one location, which is the atom's center of mass.

The electron is a sphere of radius r_e, spin frequency f_e, composed of two mathematical points that create a spinning sphere about an axis, ranging anywhere from zero to 360 degrees. This sphere spins with a speed $v_o = \alpha c$.

The proton is a sphere with spin frequency f_e/α^2, that has two mathematical points creating two spinning spheres about an axis,

73

ranging anywhere from zero to 360 degrees. These two spheres spin with a speed c/α.

The electron mass is m_e and proton mass is m_e/α^2.

The orbital frequency must be less than or equal to the spin frequency. Because this mathematical universe has an aether, all spins causes the aether to rotate.

These are the foundations of the atoms.

HYDROGEN ATOM

The electron orbits a location at a distance r_e/α and has an orbital frequency f_e, yielding an orbital speed c. The electron has a mass m_e.

The proton orbits a location at a distance αr_e and has an orbital frequency f_e/α^2, yielding an orbital speed c. The proton orbital mass is m_e/α^2.

In this mathematical universe, the relationship $E = mc^2$ and again, $E = hf$, leads to $m_{eo}c^2$. The electron mass orbiting the location is equal to hf_e with a frequency f_e. Therefore, $m_{eo} = m_e$, or the orbit creates the mass $m_{eo} = m_e$, for a total electron mass in the Hydrogen atom of $m_e(1 + 1)$.

Not only does the electron spinning at any frequency give it a certain mass. Its orbiting a particular point at any frequency creates mass as well.

The same reasoning yields a total proton mass of $2m_e/\alpha^2$. As for mass and energy, the proton's orbit creates the electron mass m_e.

The total hydrogen mass is $2m_e(1 + 1/\alpha^2)$.

In the mathematical universe, the electron mass is $m_e = hf_e/$

v_o^2. With $\lambda_e = 4.56982626\text{x}10^{-12}\text{m}$ and $f_e = 6.560262074\text{x}10^{19}\text{Hz}$, the result yield $m_e = 4.64396851\text{x}10^{-32}$ and a hydrogen mass of $1.743430509\text{x}10^{-27}$.

The electron mass plus the electron orbiting the point creates two masses. The same is true for the orbiting proton. The electron is a sphere having a spin axis of 45 degrees and the proton is a sphere having a spin of 45 degrees. The electron sphere radius is 18769 times larger than the proton sphere radius. The only sphere it can react with is the outer electron sphere, and the only reaction that can occur happens through the spin. Spheres with opposite spin directions attract, but spheres with same spin directions repel. Thus the formation of bonds is dependent only on spin directions—there are no charges, but there are zero magnitude electric vectors. Adding the proton mass, $1.67262192369\text{x}10^{-27}\text{kg}$, and electron mass, $9.1093837015\text{x}10^{-31}\text{kg}$, yields $1.673532862\text{x}10^{-27}\text{kg}$, which is quite a bit off from the mathematical hydrogen mass, $1.743430509\text{x}10^{-27}$. The reason for the difference is somewhat complex and has to do with percentages of Hydrogen's isotopes entering into the Hydrogen's mass calculation.

The next element occurs in 3 ways:

1. One electron having mass $2m_e$ and one proton having mass $2m_e/\alpha^2$ creating one sphere with a spin frequency $2f_e$.

2. Two electrons each having mass m_e and same spin, and two protons each having mass m_e/α^2 and same spin.

3. Two electrons each having mass m_e and opposite spin, and two protons each having mass m_e/α^2 and opposite spin.

The third case is the most stable and non-reactant. it is Helium. The first two are usually named Deuterium and called Hydrogen isotopes.

There is a size constraint in atom building. The obvious one is the nucleus with a proton with a radius r_e having an orbital radius αr_e cannot exist because of size constraints. The expression $[v_o i_x]$ $[(2\pi r_e/n)nf_e \cos 2\pi nf_e ti_x]$ gives the movement of the mathematical point along the i_x direction. The spherical radius is r_e/n and not the constant r_e. This transformation changes the dynamics in atom formation. The electron with frequency f_e has radius r_e, but the electron of frequency $2f_e$ has radius $r_e/2$, therefore any particle with frequency nf_e has radius r_e/n. This solution states that the spinning speed, no matter radius size, is always v_o, hence the charge is always negative. This of course is preposterous, but does charge polarity really matter if the particle orbits with speed c? The mass only depends on the spin frequency, hence a spin frequency nf_e has mass nm_e. The orbital radius αr_e may hold 137 spheres of radius $\alpha^2 r_e$. This then becomes a mechanism for how heavy particles can be packed into a nuclear radius. This mechanism obviates the strong nuclear force. When the expression keeps r_e constant it is the speed v_o/n that gets smaller as n increases.

Conventionally, Helium has two protons, two electrons and two neutrons. Taking the NIST neutron mass of $1.67492749804\times10^{-27}$ yields the calculated Helium mass 6.69692072×10^{-27}kg. The mathematical Helium atom has two opposite spins f_e/α^2 protons, two opposite spin f_e electrons, and no neutron. The mathematical Helium mass is exactly 4 times the mathematical Hydrogen mass $6.973722036\times10^{-27}$ a bit off but mathematics is simplifying the nucleus from one that convention maintains has neutrons and protons, to one just having protons. The two colliding opposite spin double mass protons do not create spinning spheres and instead creates two non spinning spheres. Since it is the spin causing attraction and repulsion, then the mathematical Helium atom cannot attract or repel any other atom, therefore it is an inert atom.

The Helium atom has two "protons," each with a mass m_e/α^2 and spin frequency f_e/α^2, and also has two electrons, each with mass m_e with spin frequency f_e. The two Helium "protons" collide and create an orbital frequency $2f_e/\alpha^2$, and the two electrons collide to create and orbital frequency $2f_e$. Although Helium and Deuterium have the same electron and proton numbers they have different masses because the two Helium electrons collide with each other in their orbital path, and the two "protons" collide with each other along their orbital path. These orbital path collisions do not occur in Deuterium and therefore has a lesser mass. This mechanism of orbital collision creates extra mass, although the number of particles in the non colliding orbiting particles are the same.

The radius r_e/n does not limit the number of chemical atoms and the possibilities are enormous. The reason we are limited is because we try to fit two comparable size nuclei and expect the result to be a comparable size nucleus, hence the creation does not last long. Civilization limits widens once we harness atom creation. A word of caution some elements may have a mass larger than the Sun, and have a comparable mass to a black hole.

The reason that stable atoms need for the number of electrons to equal the number of protons is that such equality preserves the center of mass and the constant orbital speed c. It is not that the charges have to be equal to create a zero electric field.

Nuclear physics does not exist in the mathematical universe, because there are no large number of protons packed into a nucleus and held there by the so called strong nuclear force. The protons remain in place with other protons because they orbit the atomic center of mass with speed c, thereby loosing their charge. The weak nuclear force, supposedly due to some protons and electrons colliding with each other due to orbital space constraints, does not exist either. In the mathematical universe, all that is happening is that the mathematical point is being prevented from spinning at its specific frequency.

For the lay reader this may be a rather complex section. What it shows is that the atoms can be created from the mathematical point, which unifies the two. The mathematical point is the genesis of all the elements.

For atoms, charges must orbit the atom's center of mass with the speed of light, c. That is the only way the atom can be stable. The charges' respective spin cause the charges' orbital speed. The orbital frequency cannot be greater than the spin frequency of the orbiting particle. The nucleus is just protons orbiting the atom's center of mass with speed c. There are no neutrons in the nucleus. The electron spin produces the bonding that occurs in atoms and compounds. There is no electrical static charge bond.

GRAVITY

More than four centuries ago, on March 8, 1618, after many trials and errors, Johannes Kepler formulated his third law of planetary motion: The square of the planet's orbital period, T is proportional to the cube of the planet's radial distance, R from the Sun, or

$$T^2 = kR^3,$$

where k is a constant.

To quote Kepler:

The chief aim of all investigations of the external word should be to discover the rational order and harmony which has been imposed on it by God and which He revealed to us in the language of mathematics.

One of the most striking aspects of the law is that it just has time, T^2, and space, R^3. The orbital speed $v = 2\pi R/T$ transforms $T^2 = kR^3$ to

$$v^2 = k_1/R,$$

or,

$$v = k_3/\sqrt{R},$$

where k_3 is a constant.

The equation shows that there is an acceleration towards the center of the Sun. The equation does not mention anything about mass or charge. It is independent of mass or charge. Newton assigned the cause of the acceleration to the masses attracting each other, thus making Kepler's observation $T^2 = kR^3$ more complicated and transforming it into

$$F = GMm/R^2,$$

where F is the force of attraction between mass M and mass m, R is the radial distance between the two mass centers, and G is the gravitational constant.

Newton provides the cause for the third law, but Kepler says nothing about mass or forces. All his law states is that there is a velocity gradient with zero velocity at infinity and with infinite orbital speed at zero radius. The gradient is the acceleration and the acceleration is gravity.

The question is what gives rise to Kepler's third law? A spinning sphere has the spinning speed $(4/3)\pi R^3/T$ and the speed change direction with a frequency $1/T$, or

$$T^2 = kR^3.$$

This is the equation of a spinning sphere.

When Kepler discovered that Jupiter has orbiting moons, although he was unable to detect the planet's spin, he predicted that Jupiter had a spin. He also theorized that, perhaps, magnetism caused the space to orbit Jupiter.

The law governing gravity is the density of the mathematical point of 389 points per cubic cm. At this density, the gravitational force and the outward centrifugal force are equal and cancel each other out. But when the particle density is greater than 389 particles per cubic cm, the gravitational force is greater than the centrifugal force. Any celestial body orbiting the Sun exists in an environment that has 389 particles per cubic cm. In any given cosmic location, the object exists in the sea of an infinite number of spinning stars and planets, and the object finds itself orbiting the star or planet with the shortest orbiting time period.

Applying Newton's gravitational law to the Sun, Earth, and Moon, analysis shows that the ratio of the Sun to Earth's gravitational force for the moon is about 2.19. That means the Sun attracts the Moon with a force 2.19 times stronger than the Earth's attraction for the moon, yet the Moon orbits the Earth. The reason is that Moon orbits the Earth in 27.3 days, but the Moon would orbit the Sun in about 365 days.

Two mathematical points create two spinning spheres combining to form one spinning sphere about an axis. The spinning sphere causes the universe to orbit the spinning sphere. This spinning sphere with the equation $(4/3)\pi R^3/T^2 = k$ causes the universe to orbit it exactly according to Kepler's third law of motion $T^2 = kR^3$. The astonishing conclusion is that this equation creates a velocity gradient throughout the entire universe, and this velocity gradient is gravity.

Considering all the attention gravity has received, including its role in the formulation of general relativity, it is not a fundamental force in the universe. Its origin is simply due to the spin of the interaction between two mathematical points. Strange as it may seem, there is no gravitational constant, no mass attracting another mass. What exists is a spinning object causing a velocity gradient throughout the cosmos.

Finally, the new mathematical point unifies the electromagnetic forces and the gravitational forces. Due to the creation of mass through the mathematical point spin, there is no strong or weak nuclear force. Therefore, the new mathematical point unifies the universe.

IN CONCLUSION

The challenge is for science to realize that mathematics has to change from the abstract Euclidian point, "that which has no part," to a point that is real and has volume. No matter how infinite the universe is, it has a volume with Cartesian coordinates that identify all locations in the universe. Because the point occupies space, no two points can be in the same place at the same time. These are the axioms. Because the point is real, or has a volume, the point has an interior, which allows for further investigation. This is not possible in an Euclidian point which has no dimension.

Point movement outlines a continuous mathematical function, and constant speed point movement creates a uniform mathematical universe. Because the mathematical domain of all functions is from minus to positive infinity, the mathematical point can neither be created or destroyed. Taking the point with a volume moving with a constant speed, and examining its surface *and* interior gives rise to a fundamental change in the way we view the universe. Because linear motion now occurs with the moving point rotating about its axis of motion, Newton's first law of motion must be adjusted accordingly. Point rotation is spin, which is a moving magnetic vector that creates an electric vector and produces a charge, along with Maxwell's electromagnetic equations.

Taking this further, the electric vector, a consequence of motion, changes polarity depending on the point's speed. The point is negative, neutral, or positive depending on its speed being less, equal, or greater than the speed of light. When the point travels at the speed of light, the point has no electromagnetic quality. This is a necessary condition for atom creation.

Newton's second law of motion and object spin results in mass. Motion is impossible without resistance. Since everything in the universe is moving, there must be resistance, and the resistance is the aether. There is no need to do experiments to show this, because motion is impossible without resistance.

The fact that motion creates charge and mass is a startling discovery. But mass creation has been here all along. Another incredible finding is that spin creates heat without any external heat source acting on the moving point. Point collision also creates heat, therefore constantly colliding points create heat, and the collision heat creates a temperature close to the temperature found in the cosmic microwave background radiation (CMB). This leads to the conclusion that photons are the mathematical points that create the reality of the universe.

The mathematical point's spin and collision create thermodynamics. Heat is simply due to spin and collision, and has nothing to do with nuclear reactions. Stars produce heat simply because they spin, not because of nuclear reactions.

Rapidly spinning particles are the sole source of the star's heat, not the nuclear reactions that may or may not be taking place. Spin creates the heat in the spinning particle. With a high enough spinning frequency, heat can be greater than the Sun's heat. Spinning at a lower frequency creates planets and creates the heat in the planets. It is hard to imagine that the Earth's core is undergoing constant nuclear reactions.

Point collision also creates point harmonic vibrations, which are the frequencies and sinusoidal functions existing in the universe. These frequencies and sinusoids are the cause for the quantum observations.

The sinusoids leads to circle creation, which engenders electrons, neutrons, protons, planets, stars and black holes. The last are stars that spin so fast that they are beyond the optical spectrum. They appear dark, but they are as real as the other visible stars. As they spin, electrons and protons interact to form the atoms. These electrons and protons with variable radii interact to form the atoms. The variable radius, r_e/n, in the electron and proton explains how the nucleus radius stays constant despite packing in more and more protons, and gets rid of the strong nuclear force. This variable radius opens the portal for a large number of stable element creation. At any given time, only two point collisions preserve the mathematical universe, which explains the universal observation of planar orbits.

Finally, spin explains the genesis of gravity according to Kepler's third law of planetary motion, which shows that gravity is the same as the equation of spherical spin.

If science is to advance, this essay is making the plea to change the mathematical point from "that which has no part" to one that includes volume in the equation. Throughout this mathematical analysis, a basic message keeps coming back: That the only reality aspect is space and time. Einstein was correct in his contention that everything in the universe is a consequence of the interaction between space and time.

In making the change from "that which has no part" to a three-dimensional point will lead to the discovery that the new mathematics unifies the universe. This understanding leads to a host of surprising implications for those schooled in the prevailing, non-dimensional model. Here are a few:

The idea that the Earth's spin creates different speed as a location varies from equator to poles is erroneous; what varies is the time while the speed stays constant. The force throughout the spin is constant as well, but the circumference of the Earth varies; hence there is more heat generated at the equator than at the poles. Thus, there is no need to fire a rocket ship from the equator rather than a higher or a lower latitude.

Given Kepler's third law, $T^2 = kR^3$, is the spherical spin equation, which describes an acceleration and therefore gravity. By implication, there is no Gravitational constant, G, and mass does not create gravity. The search for gravitational waves is futile, a waste of time. There are none.

The new understanding of Kepler's third law provides an reasonable explanation of orbits, the solar system and the galaxies. It accounts for dark matter as missing energy coming from black holes. Consider if for some reason we could not detect our Sun but could observe the planets. We would conclude that more than 99 percent of the solar system's energy is missing. In a similar manner, we cannot directly observe a black hole and its missing energy. But we can end all speculation and search for dark energy and dark matter by simply conceiving of a black hole as a star.

Another discovery is that the atom is not made up of same size particles. The equation $[(4/3)\pi r_o^2 v_o][2\pi(r_e/n)nf_e\cos(2\pi nf_e t]$ implies that the radius of the sphere formation is r_e/n, which decreases as n increases. But because the mass only depends on the frequency nf_e, the mass increases as n increases. The speed of the formed sphere is v_o. The charge of the formed spheres is always the same negative charge. The nucleus then is always negative, which is quite a radical departure from our current understanding of the atom.

More importantly, this varying radius opens a portal to the creation of almost an infinite number of stable elements.

Given that mass is directly proportional to frequency, particles can take on any mass by spinning at a higher frequency. There is no need to have 10 electrons each having a radius r_e and frequency f_e when there can be one electron with frequency $10f_e$, and radius $r_e/10$. This decrease in spherical size is new in the universe and makes the strong nuclear force unnecessary. Positing $E = hf = mc^2 = kT/\lambda$ explains the universe better than any current theory.

The new, three-dimensional, mathematical point simply cannot be ignored. It determines the movement of a real object and analyzes the interaction of the object's movement and vectors. From there, we can derive electromagnetism, mass, charge, an aether, and heat. By adding collision, we get electrons, neutrons, and protons. They, in turn, generate black holes, stars, planets, moons, and asteroids.

What an exciting world of possibilities opens up for all of us!

APPENDIX
VECTOR MULTIPLICATION

In vector multiplication, there are dot ($\cdot$) product and cross (x) product multiplication. The dot product of two vectors will always lead to a scalar result (no vectors), while the cross product will always yield a vector result.

For example, in the case of:

With VECTOR $\mathbf{X} = Xi_x + Yi_y + Zi_z$ and VECTOR $\mathbf{A} = Ai_x + Bi_y + Ci_z$, taking the vector dot product will yield the scalar quantity

$$\mathbf{X} \cdot \mathbf{A} = XA + YB + ZC$$

with the rules $i_x \cdot i_x = 1$, $i_y \cdot i_y = 1$, $i_z \cdot i_z = 1$,
but $i_x \cdot i_y = 0$, $i_x \cdot i_z = 0$, $i_y \cdot i_z = 0$.

In the case of the vector cross product

$$\mathbf{X} \times \mathbf{A} = (YC - ZB)i_x + (ZA - XC)i_y + (XB - YA)i_z$$

with the rules

$i_x \times i_y = i_z$, but $i_y \times i_x = -i_z$,
$i_y \times i_z = i_x$, but $i_z \times i_y = -i_x$,
$i_z \times i_x = i_y$, but $i_x \times i_z = -i_y$,
and $i_x \times i_x = 0$, $i_y \times i_y = 0$, and $i_z \times i_z = 0$.

In the vector cross product, the order is very important.

For the values X, Y, Z, A, B, C, the polarity for multiplication holds. That is, -A times a positive X is -AX and -A times a -X is AX, or multiplication of two like signs is positive and multiplication of two unlike signs is negative.

ACKNOWLEDGMENTS

I have been fortunate to encounter many people in my life who have influenced me profoundly. I wish to thank them all, and many know who they are.

Here I want to mention and thank just a few specifically:

- My son, Ian, who was the original inspiration for this quest. He has brought into this world my most enjoyable grandson Bair, who might just give me the impetus to explain the universe once again when he becomes old enough to follow my ruminations.

- Hope Estepan, my wife, for being such a steadfast supporter.

- Chris Angermann, who continues to profess his inability to follow some of the more advanced math, but has made the essay more readable nontheless.

www.ingramcontent.com/pod-product-compliance
Lightning Source LLC
Chambersburg PA
CBHW031813190326
41518CB00006B/325